Textbook of Bioinformatics

Textbook of Bioinformatics

Regan Knight

STATES
ACADEMIC PRESS
www.statesacademicpress.com

States Academic Press,
109 South 5th Street,
Brooklyn, NY 11249, USA

Visit us on the World Wide Web at:
www.statesacademicpress.com

ISBN: 978-1-63989-510-6 (Hardback)

Cataloging-in-Publication Data

Textbook of bioinformatics / Regan Knight.
p. cm.
Includes bibliographical references and index.
ISBN 978-1-63989-510-6
1. Bioinformatics. 2. Biology--Data processing. 3. Computational biology. I. Knight, Regan.
QH324.2 .B56 2022
570.285--dc23

TABLE OF CONTENTS

PREFACE

This book is a culmination of my many years of practice in this field. I attribute the success of this book to my support group. I would like to thank my parents who have showered me with unconditional love and support and my peers and professors for their constant guidance.

Bioinformatics is a field which aims to understand biological data. It is concerned with developing diverse software tools and methods which can facilitate the comprehension of complex biological data. It is an interdisciplinary field that combines concepts and knowledge from the areas of computer science, mathematics, biology, statistics and information engineering. These areas help in an advanced analysis and interpretation of data. The most prominent applications of bioinformatics include identifying the genetic basis of a disease, understanding differences between populations, identification of single nucleotide polymorphisms, understanding protein sequences, identifying candidate genes, understanding the desirable properties amongst agricultural species, etc. This field has contributed in text mining of biological literature for developing gene and biological ontologies. This book provides comprehensive insights into bioinformatics. It elucidates new techniques and their applications in a multidisciplinary manner. This textbook is a complete source of knowledge on the present status of this important field.

The details of chapters are provided below for a progressive learning:

Chapter - What is Bioinformatics?

Bioinformatics is an interdisciplinary field of science that develops software tools and methods for analyzing and interpreting biological data. It deals with sequence analysis such as DNA sequencing, comparative genomics analysis, etc. This is an introductory chapter which will introduce briefly all the significant aspects of bioinformatics.

Chapter - Sequence Alignment and Analysis

Sequence alignment is termed as the arrangement of DNA and RNA sequences for identifying functional and structural relationships between the sequences. The process of subjecting a DNA, RNA or peptide sequence to any analytical method is referred to as sequence analysis. This chapter discusses the subject of sequence alignment and analysis in detail.

Chapter - Phylogenetic Analysis

The study of evolutionary history and their relationships among individuals and group of organisms is called phylogenetics. Such estimation of evolutionary relationships among biological species is done through an analysis known as phylogenetic analysis. The topics elaborated in this chapter will help in gaining a better perspective about phylogenetic analysis.

Chapter - Biological Database

Biological database refers to the stores of biological information. It includes genomics, proteomics, metabolomics, microarray gene expression, gene function and mutation. This chapter has been carefully written to provide an easy understanding of the related concepts of biological database.

Chapter - Bioinformatics Tools and Software

There are various tools and software that are used in bioinformatics. Some of them are Clustal, PHYLIP, FASTA, BLAST, BLAT, BioSLAX, BioPerl, BioJava, ArrayTrack, etc. This chapter closely examines the different bioinformatics tools and software to provide an extensive understanding of the subject.

Regan Knight

What is Bioinformatics?

CHAPTER 1

Bioinformatics is an interdisciplinary field of science that develops software tools and methods for analyzing and interpreting biological data. It deals with sequence analysis such as DNA sequencing, comparative genomics analysis, etc. This is an introductory chapter which will introduce briefly all the significant aspects of bioinformatics.

Bioinformatics is the science of storing, retrieving and analysing large amounts of biological information. It is a highly interdisciplinary field involving many different types of specialists, including biologists, molecular life scientists, computer scientists and mathematicians. The term bioinformatics was coined by Paulien Hogeweg and Ben Hesper to describe "the study of informatic processes in biotic systems" and it found early use when the first biological sequence data began to be shared. Whilst the initial analysis methods are still fundamental to many large-scale experiments in the molecular life sciences, nowadays bioinformatics is considered to be a much broader discipline, encompassing modelling and image analysis in addition to the classical methods used for comparison of linear sequences or three-dimensional structures.

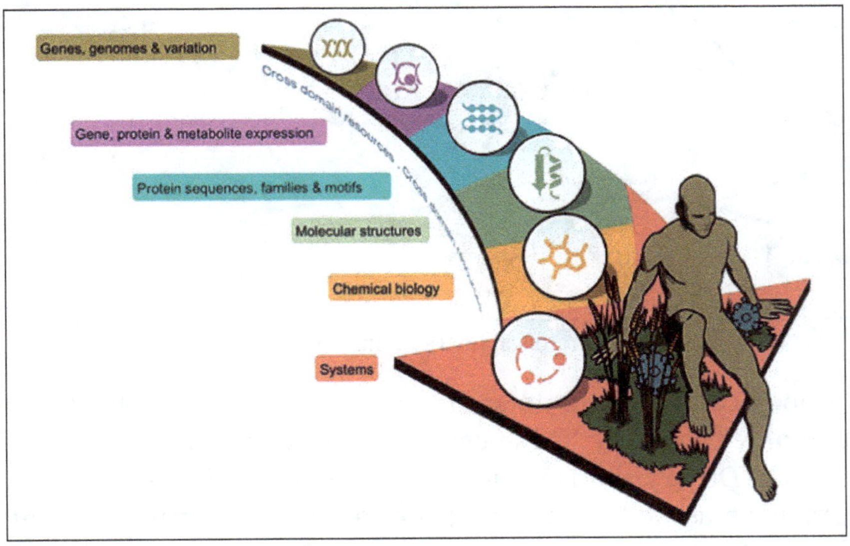

A broad overview of the different types of data that fall within the scope of bioinformatics. Traditionally, bioinformatics was used to describe the science of storing and analysing biomolecular sequence data, but the term is now used much more broadly, encompassing computational structural biology, chemical biology and systems biology (both data integration and the modelling of systems).

Distinction from Medical Informatics

Bioinformatics is distinct from medical informatics: The interdisciplinary study of the design, development, adoption and application of IT-based innovations in healthcare services delivery, management and planning. Somewhere in between the two disciplines lies biomedical informatics: The interdisciplinary field that studies and pursues the effective uses of biomedical data, information, and knowledge for scientific enquiry, problem solving and decision making, motivated by efforts to improve human health.

Recently initiated projects, such as the 100,000 Genomes Project, are bridging the gaps between these disciplines, but on the whole bioinformatics deals with research data and uses it for research purposes, medical informatics deals with data from individual patients for the purposes of clinical management, (diagnosis, treatment, prevention) and biomedical informatics attempts to bridge these two extremes.

Bioinformatics develops and applies computational methods to analyse large collections of biological data, such as genetic sequences, cell populations or protein samples, to make new predictions or discover new biology. The computational methods used include analytical methods, mathematical modelling and simulation.

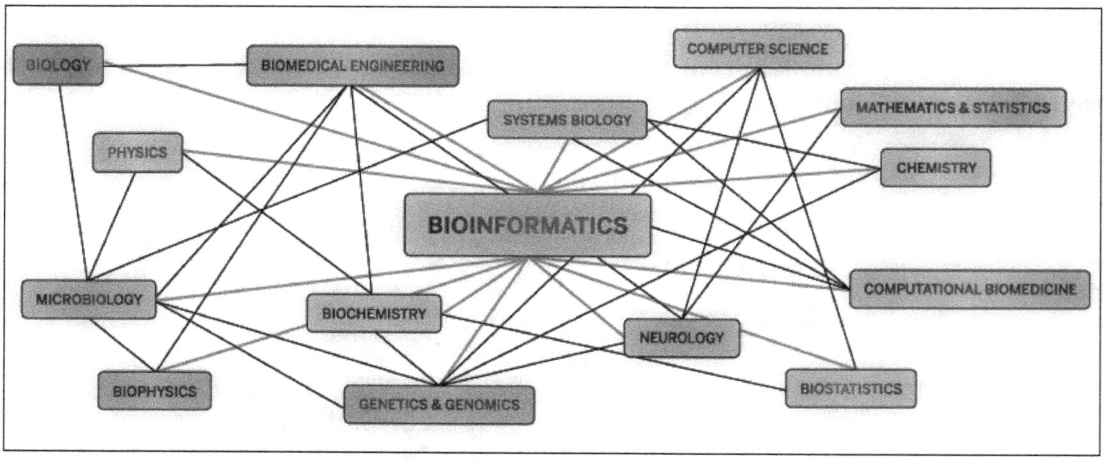

Interdisciplinary Bioinformatics.

Our human body can be break down into small machineries of cells which is involved in complex processes. These cells are controlled by the central processing unit called DNA (De-oxyribo Nucleic Acid). Understanding DNA can reveal a lot about the organism as well as the chances of diseases in future. Current technologies including NGS (Next Generation Sequencing) has generated large amount of data. These big data (Genome, Transcriptome, Proteome and Metabolome) should be organised into databases and must be analysed. The outcomes of analysis these large data (termed as Big Data) are utilized in healthcare, preventive medicine and drug discovery.

COMPONENTS OF BIOINFORMATICS

Technology and Computing Power

Technology is the collection of techniques, skills, methods, and processes applied to make life easier. Technology can be the knowledge of techniques, processes. Computing power is the ability to undertake or be used for computation. Specifically the ability of a computer to perform work, often considered in terms of the number of instructions that can be carried out in a given time, or with reference to the amount of random access memory present.

Creation of Databases

This involves the organizing, storage and management the biological data sets. The databases are accessible to researchers to know the existing information and submit new entries, e.g. protein sequence data bank for molecular structure. Databases will be of no use until analysed.

Development of Algorithms and Statistics

This involves the development of tools and resources to determine the relationship among the members of large data sets e.g. comparison of protein sequence data with the already existing protein sequences.

Analysis of Data and Interpretation

The appropriate use of components to analyse the data and interpret the results in a biologically meaningful manner. This includes DNA, RNA and protein sequences, protein structure, gene expression profiles and biochemical pathways.

Computational Biology

Broadly speaking, computational biology is the application of computer science, statistics, and mathematics to problems in biology. Computational biology spans a wide range of fields within biology, including genomics/genetics, biophysics, cell biology, biochemistry, and evolution. Likewise, it makes use of tools and techniques from many different quantitative fields, including algorithm design, machine learning, Bayesian and frequentist statistics, and statistical physics.

Molecular Biology

Molecular biology is the study of biology at a molecular level. The field overlaps with other areas of biology and chemistry, particularly genetics and biochemistry. Molecular biology chiefly concerns itself with understanding the interactions between the various systems of a cell, including the interrelationship of DNA, RNA and protein synthesis and learning how these interactions are regulated. Researchers in molecular biology use specific techniques native to molecular biology, but increasingly combine these with techniques and ideas from genetics and biochemistry. There is not a hard-line between these disciplines as there once was. Molecular biology is the study of molecular underpinnings of the process of replication, transcription and translation of the genetic material. The central dogma of molecular biology where genetic material is transcribed into RNA and then translated into protein, despite being an oversimplified picture of molecular biology, still provides a good starting point for understanding the field. Much of the work in molecular biology is quantitative, and recently much work has been done at the interface of molecular biology and computer science in bioinformatics and computational biology.

Genomics

Genomics is an interdisciplinary field of science within the field of molecular biology. A genome is a complete set of DNA within a single cell of an organism, and as such, focuses on the structure, function, evolution, and mapping of genomes. Genomics aims at the collective characterization and quantification of genes, which direct the production of proteins with the assistance of enzymes and messenger molecules. Genomics also involves the sequencing and analysis of genomes. Advances in genomics have triggered a revolution in discovery-based research to understand even the most currently complex biological systems such as the brain. In contrast to genetics, which refers to the study of individual genes and their roles in inheritance, genomics uses high throughput DNA sequencing and bioinformatics to assemble, and analyze the function and structure of entire genomes.

Computer Sciences

The study of the theory, experimentation, and engineering that form the basis for the design and use of computers. It is the scientific and practical approach to computation and its applications and the systematic study of the feasibility, structure, expression, and mechanization of the methodical procedures (or algorithms) that underlie the acquisition, representation, processing, storage, communication of, and access to information. An alternate, more succinct definition of computer science is the study of automating algorithmic processes that scale. A computer scientist specializes in the theory of computation and the design of computational systems. Its fields can be divided into a variety of theoretical and disciplines. Some fields, such as computational complexity theory (which explores the fundamental properties of computational and intractable problems), are highly abstract, while fields such as computer graphics emphasize real-world visual applications. Other fields still focus on challenges in implementing computation. For example, programming language theory considers various approaches to the description of computation, while the study of computer programming itself investigates various aspects of the use of programming language and complex systems. Human–computer interaction considers the challenges in making computers and computations useful, usable, and universally accessible to humans.

Bioinformatics brings together large data bases of biological information and computational techniques of analysis.

DATA IN BIOINFORMATICS

The classic data of bioinformatics include DNA sequences of genes or full genomes; amino acid sequences of proteins; and three-dimensional structures of proteins, nucleic acids and protein–nucleic acid complexes. Additional "-omics" data streams include: transcriptomics, the pattern of RNA synthesis from DNA; proteomics, the distribution of proteins in cells; interactomics, the patterns of protein-protein and protein–nucleic acid interactions; and metabolomics, the nature and traffic patterns of transformations of small molecules by the biochemical pathways active in cells. In each case there is interest in obtaining comprehensive, accurate data for particular cell types and in identifying patterns of variation within the data. For example, data may fluctuate depending on cell type, timing of data collection (during the cell cycle, or diurnal, seasonal, or annual variations), developmental stage, and various external conditions. Metagenomics and metaproteomics extend these measurements to a comprehensive description of the organisms in an environmental sample, such as in a bucket of ocean water or in a soil sample.

Bioinformatics has been driven by the great acceleration in data-generation processes in biology. Genome sequencing methods show perhaps the most dramatic effects. In 1999 the nucleic acid sequence archives contained a total of 3.5 billion nucleotides, slightly more than the length of a single human genome; a decade later they contained more than 283 billion nucleotides, the length of about 95 human genomes. The U.S. National Institutes of Health has challenged researchers by setting a goal to reduce the cost of sequencing a human genome to $1,000; this would make DNA sequencing a more affordable and practical tool for U.S. hospitals and clinics, enabling it to become a standard component of diagnosis.

Storage and Retrieval of Data

In bioinformatics, data banks are used to store and organize data. Many of these entities collect DNA and RNA sequences from scientific papers and genome projects. Many databases are in the hands of international consortia. For example, an advisory committee made up of members of the European Molecular Biology Laboratory Nucleotide Sequence Database (EMBL-Bank) in the United Kingdom, the DNA Data Bank of Japan (DDBJ), and GenBank of the National Center for Biotechnology Information (NCBI) in the United States oversees the International Nucleotide Sequence Database Collaboration (INSDC). (Similar conditions apply to nucleic acid and protein structures). There also exist genome browsers, databases that bring together all the available genomic and molecular information about a particular species.

The major database of biological macromolecular structure is the worldwide Protein Data Bank (wwPDB), a joint effort of the Research Collaboratory for Structural Bioinformatics (RCSB) in the United States, the Protein Data Bank Europe (PDBe) at the European Bioinformatics Institute in the United Kingdom, and the Protein Data Bank Japan at Ōsaka University. The homepages of the wwPDB partners contain links to the data files themselves, to expository and tutorial material (including news items), to facilities for deposition of new entries, and to specialized search software for retrieving structures.

Information retrieval from the data archives utilizes standard tools for identification of data items by keyword; for instance, one can type "aardvark myoglobin" into Google and retrieve the molecule's amino acid sequence. Other algorithms search data banks to detect similarities between data items. For example, a standard problem is to probe a sequence database with a gene or protein sequence of interest in order to detect entities with similar sequences.

GOALS OF BIOINFORMATICS

The development of efficient algorithms for measuring sequence similarity is an important goal of bioinformatics. The Needleman-Wunsch algorithm, which is based on dynamic programming, guarantees finding the optimal alignment of pairs of sequences. This algorithm essentially divides a large problem (the full sequence) into a series of smaller problems (short sequence segments) and uses the solutions of the smaller problems to construct a solution to the large problem. Similarities in sequences are scored in a matrix, and the algorithm allows for the detection of gaps in sequence alignment.

Although the Needleman-Wunsch algorithm is effective, it is too slow for probing a large sequence database. Therefore, much attention has been given to finding fast information-retrieval algorithms that can deal with the vast amounts of data in the archives. An example is the program BLAST (Basic Local Alignment Search Tool). A development of BLAST, known as position-specific iterated- (or PSI-) BLAST, makes use of patterns of conservation in related sequences and combines the high speed of BLAST with very high sensitivity to find related sequences.

Another goal of bioinformatics is the extension of experimental data by predictions. A fundamental goal of computational biology is the prediction of protein structure from an amino acid sequence. The spontaneous folding of proteins shows that this should be possible. Progress in the development of methods to predict protein folding is measured by biennial Critical Assessment of Structure Prediction (CASP) programs, which involve blind tests of structure prediction methods.

Bioinformatics is also used to predict interactions between proteins, given individual structures of the partners. This is known as the "docking problem." Protein-protein complexes show good complementarity in surface shape and polarity and are stabilized largely by weak interactions, such as burial of hydrophobic surface, hydrogen bonds, and van der Waals forces. Computer programs simulate these interactions to predict the optimal spatial relationship between binding partners. A particular challenge, one that could have important therapeutic applications, is to design an antibody that binds with high affinity to a target protein.

Initially, much bioinformatics research has had a relatively narrow focus, concentrating on devising algorithms for analyzing particular types of data, such as gene sequences or protein structures. Now, however, the goals of bioinformatics are integrative and are aimed at figuring out how combinations of different types of data can be used to understand natural phenomena, including organisms and disease.

References

- Bioinformatics, science: britannica.com, Retrieved 12 March, 2020
- What-bioinformatics, training, bioinformatics-terrified-0: ebi.ac.uk, Retrieved 13 August, 2020
- What-is-bioinformatics-703170763999: towardsdatascience.com, Retrieved 25 April, 2020
- Dahiya BL, Lata M. Bioinformatics impacts on medicine, microbial genome and agriculture. Journal of Pharmacognosy and Phytochemistry. 2017; 6(4):1938- 1942
- Goals-of-bioinformatics, bioinformatics, science: britannica.com, Retrieved 08 February, 2020

Sequence Alignment and Analysis

CHAPTER 2

Sequence alignment is termed as the arrangement of DNA and RNA sequences for identifying functional and structural relationships between the sequences. The process of subjecting a DNA, RNA or peptide sequence to any analytical method is referred to as sequence analysis. This chapter discusses the subject of sequence alignment and analysis in detail.

The alignment of biological sequences is probably the most important and most accomplished in the field of bioinformatics. Their ultimate goal is to determine the similarity between different sequences. The sequence alignment is therefore a great number of applications:

- Gene Finding: Having identified all ORFs of a genome can be used their lengths to estimate the probability of actually constituting genes, these are ab initio methods mentioned before. On the other hand, if there is a gene from another organism with a good alignment with the ORF found, there is strong evidence that this ORF is a gene.
- Function Prediction: The alignment of sequences to determine if two genes are similar. That way if the first function of these genes is known we can assign the same function as the second.
- Genome Sequence Assembly: The de-novo assembly is a very important task in bioinformatics is based on the alignments between short DNA sequences obtained by the new-generation sequencers.

One of the main applications of sequence alignment is the identification of homologous genes. Living organisms share a large number of genes descended from common ancestors and have been maintained in different organisms due to its functionality but accumulate differences that have diverged from each other. These differences may be due to mutations that change a symbol (nucleotide or amino acid) for another or insertions /deletions, indels, which insert or delete a symbol in the corresponding sequence.

Two genes are homologous if they share a common ancestor. Therefore, the homology is a dichotomous characteristic, i.e., given two genes are either homologous genes or not. However, given two sequences corresponding to two genes, can be said that there are different levels of similarity based on an alignment between them. The key task is to determine whether a good alignment between two sequences is significant enough to consider that both genes are homologous.

There are two different forms of homology. When the origin of two homologous genes is due to a process of gene duplication within the same species these genes are called paralogs genes, whereas when the origin is due to a speciation process resulting in homologous genes in these different species are called orthologous genes.

The first step to compare two sequences is, usually, to align them.

```
No alignment

 CGATGCTAGCGTATCGTAGTCTATCGTAC
       |   ||
 ACGATGCTAGCGTTTCGTATCATCGTA

Alignned

 -CGATGCTAGCGTATCGTAGTCTATCGTAC
 |||||||||||| ||||||||||||||
 ACGATGCTAGCGTTTCGTA-TC-ATCGTA-
```

There could be substitutions, changes of one residue with another, or gaps. Gaps are missing residues and could be due to a deletion in one sequence or an insertion in the other sequence.

Gaps complicate the alignments. Algorithms should take into account the possibility of introducing gaps and once we allow them to create gaps several alignments can be constructed between two sequences.

```
No gaps (10 matches)

 a: ATATTGCTACGTATATCAT
       ||||||||||
 b: ATATATGCTACGTATCAT

With one gap (14 matches)

 a: ATAT-TGCTACGTATATCAT
   |||| ||||||||||
 b: ATATATGCTACGTATCAT
```

```
With two gaps (16 matches)
 a: ATAT-TGCTACGTATATCAT
    |||| |||||||| ||||||
 b: ATATATGCTACG--TATCAT
```

The objective of a sequence alignment is, usually, to align the homologous positions of the two sequences. The homologous positions are the ones that come from the same position in the ancestral sequence. We don't know the ancestral sequence, so we won't be completely sure that we have succeeded. Another complementary objective could be to align protein regions that have the same structure or function.

Aligning similar sequences by any algorithm usually creates alignments that are usually correct, but when sequences are very different aligning them could be a challenge. Once a long time has passed since the split of the species the sequences can be so changed by the mutations that any meaningful similarities could have been lost an creating a meaningful alignment could be very difficult.

Evaluating the Alignments

To be able to compare the different possible alignments we can score them. We can create a scoring system that gives more points to alignments that are biologically more reasonable. Ideally we would create a scoring system that gives more points to the alignments that align the homologous positions.

A naive scoring system could be to count the number of matching positions, or the number of matching positions along 100 residues. Usually the scoring systems also take into account the number of gaps. They penalize the alignments depending on the number and the length of the gaps present. So the main features taken into account to create a scoring system are usually:

- Number of matching residues (taking into account the similarity if they are aminoacids).
- Number of mismatches.
- Number of gaps.
- Length of the gaps.

We can devise different scoring schemes with those measures. For instances:

- Scoring schema 1: match +1, mismatch: 0, gap creation: -1 gap extension: -1.
- Scoring schema 1: match +1, mismatch: -1, gap creation: -1 gap extension: 0.

Of course, one alignment will have a different score under different scoring schemes. Speaking of the score of an alignment it is meaningless if we do not take into account the scoring schema used. It also has no sense to compare the scores of different alignments done under different schemes.

Once we have decided which scoring schema to use, the alignment algorithm should try to create the alignment that obtains the maximum score under that particular scoring schema.

Every software implementation of an algorithm will usually have some default values for its parameters. These default values have been calibrated by the software creator to work well in a particular problem. The bioinformatician should be aware of how well those values apply to the particular problem at hand. Usually, if the problem is similar to the one that motivated the creation of the software the default value will work OK because the original creator of the software usually knows how to optimize his software for that task. When our problem is different from the original one we have to be aware of the changes to adapt the software to our needs.

SEQUENTIAL PATTERN MINING

Sequential pattern mining is a topic of data mining concerned with finding statistically relevant patterns between data examples where the values are delivered in a sequence. It is usually presumed that the values are discrete, and thus time series mining is closely related, but usually considered a different activity. Sequential pattern mining is a special case of structured data mining.

There are several key traditional computational problems addressed within this field. These include building efficient databases and indexes for sequence information, extracting the frequently occurring patterns, comparing sequences for similarity, and recovering missing sequence members. In general, sequence mining problems can be classified as string mining which is typically based on string processing algorithms and itemset mining which is typically based on association rule learning. Local process models extend sequential pattern mining to more complex patterns that can include (exclusive) choices, loops, and concurrency constructs in addition to the sequential ordering construct.

String Mining

String mining typically deals with a limited alphabet for items that appear in a sequence, but the sequence itself may be typically very long. Examples of an alphabet can be those in the ASCII character set used in natural language text, nucleotide bases 'A', 'G', 'C' and 'T' in DNA sequences, or amino acids for protein sequences. In biology applications analysis of the arrangement of the alphabet in strings can be used to examine gene and protein sequences to determine their properties. Knowing the sequence of letters of a DNA or a protein is not an ultimate goal in itself. Rather, the major task is to understand the sequence, in terms of its structure and biological function. This is typically achieved first by identifying individual regions or structural units within each sequence and then assigning a function to each structural unit. In many cases this requires comparing a given sequence with previously studied ones. The comparison between the strings becomes complicated when insertions, deletions and mutations occur in a string.

A survey and taxonomy of the key algorithms for sequence comparison for bioinformatics is presented by Abouelhoda and Ghanem, which include:

- Repeat-related problems: That deal with operations on single sequences and can be based on exact string matching or approximate string matching methods for finding dispersed fixed length and maximal length repeats, finding tandem repeats, and finding unique subsequences and missing (un-spelled) subsequences.
- Alignment problems: That deal with comparison between strings by first aligning one or more sequences; examples of popular methods include BLAST for comparing a single sequence with multiple sequences in a database, and ClustalW for multiple alignments. Alignment algorithms can be based on either exact or approximate methods, and can also be classified as global alignments, semi-global alignments and local alignment.

Itemset Mining

Some problems in sequence mining lend themselves discovering frequent itemsets and the order they appear, for example, one is seeking rules of the form "if a {customer buys a car}, he or she is likely to {buy insurance} within 1 week", or in the context of stock prices, "if {Nokia up and Ericsson up}, it is likely that {Motorola up and Samsung up} within 2 days". Traditionally, itemset mining is used in marketing applications for discovering regularities between frequently co-occurring items in large transactions. For example, by analysing transactions of customer shopping baskets in a supermarket, one can produce a rule which reads "if a customer buys onions and potatoes together, he or she is likely to also buy hamburger meat in the same transaction".

A survey and taxonomy of the key algorithms for item set mining is presented by Han et al. The two common techniques that are applied to sequence databases for frequent itemset mining are the influential apriori algorithm and the more-recent FP-growth technique.

Applications

With a great variation of products and user buying behaviors, shelf on which products are being displayed is one of the most important resources in retail environment. Retailers can not only increase their profit but, also decrease cost by proper management of shelf space allocation and products display. To solve this problem, George and Binu have proposed an approach to mine user buying patterns using PrefixSpan algorithm and place the products on shelves based on the order of mined purchasing patterns.

Algorithms

Commonly used algorithms include:

- GSP algorithm.
- Sequential PAttern Discovery using Equivalence classes (SPADE).
- FreeSpan.
- PrefixSpan.
- MAPres.

PAIRWISE ALIGNMENT

Pairwise sequence alignment methods are used to find the best-matching piecewise (local or global) alignments of two query sequences. Pairwise alignments can only be used between two sequences at a time, but they are efficient to calculate and are often used for methods that do not require extreme precision (such as searching a database for sequences with high similarity to a query). The three primary methods of producing pairwise alignments are dot-matrix methods, dynamic programming, and word methods; however, multiple sequence alignment techniques can also align pairs of sequences. Although each method has its individual strengths and weaknesses, all three pairwise methods have difficulty with highly repetitive sequences of low information content - especially where the number of repetitions differ in the two sequences to be aligned. One way of quantifying the utility of a given pairwise alignment is the 'maximum unique match' (MUM), or the longest subsequence that occurs in both query sequences. Longer MUM sequences typically reflect closer relatedness.

Dot-matrix Methods

The dot-matrix approach, which implicitly produces a family of alignments for individual sequence regions, is qualitative and conceptually simple, though time-consuming to analyze on a large scale. In the absence of noise, it can be easy to visually identify certain sequence features—such as insertions, deletions, repeats, or inverted repeats—from a dot-matrix plot. To construct a dot-matrix plot, the two sequences are written along the top row and leftmost column of a two-dimensional matrix and a dot is placed at any point where the characters in the appropriate columns match—this is a typical recurrence plot. Some implementations vary the size or intensity of the dot depending on the degree of similarity of the two characters, to accommodate conservative substitutions. The dot plots of very closely related sequences will appear as a single line along the matrix's main diagonal

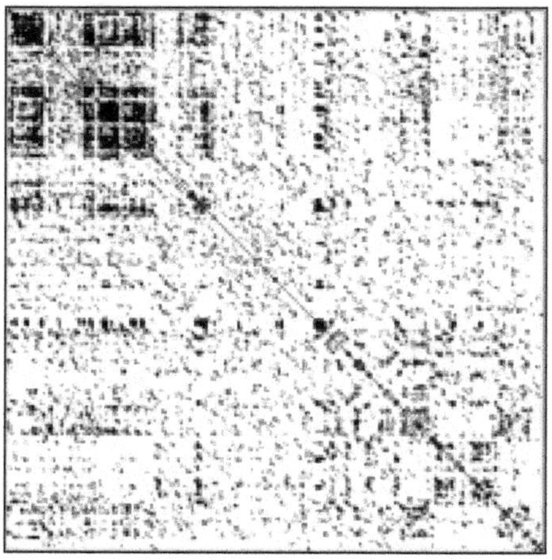

A DNA dot plot of a human zinc finger transcription factor, showing regional self-similarity. The main diagonal represents the sequence's alignment with itself; lines off the main diagonal represent similar or repetitive patterns within the sequence. This is a typical example of a recurrence plot.

Problems with dot plots as an information display technique include: noise, lack of clarity, non-intuitiveness, difficulty extracting match summary statistics and match positions on the two sequences. There is also much wasted space where the match data is inherently duplicated across the diagonal and most of the actual area of the plot is taken up by either empty space or noise, and, finally, dot-plots are limited to two sequences. None of these limitations apply to Miropeats alignment diagrams but they have their own particular flaws.

Dot plots can also be used to assess repetitiveness in a single sequence. A sequence can be plotted against itself and regions that share significant similarities will appear as lines off the main diagonal. This effect can occur when a protein consists of multiple similar structural domains.

Dynamic Programming

The technique of dynamic programming can be applied to produce global alignments via the Needleman-Wunsch algorithm, and local alignments via the Smith-Waterman algorithm. In typical usage, protein alignments use a substitution matrix to assign scores to amino-acid matches or mismatches, and a gap penalty for matching an amino acid in one sequence to a gap in the other. DNA and RNA alignments may use a scoring matrix, but in practice often simply assign a positive match score, a negative mismatch score, and a negative gap penalty. (In standard dynamic programming, the score of each amino acid position is independent of the identity of its neighbors, and therefore base stacking effects are not taken into account. However, it is possible to account for such effects by modifying the algorithm). A common extension to standard linear gap costs, is the usage of two different gap penalties for opening a gap and for extending a gap. Typically the former is much larger than the latter, e.g. -10 for gap open and -2 for gap extension. Thus, the number of gaps in an alignment is usually reduced and residues and gaps are kept together, which typically makes more biological sense. The Gotoh algorithm implements affine gap costs by using three matrices.

Dynamic programming can be useful in aligning nucleotide to protein sequences, a task complicated by the need to take into account frameshift mutations (usually insertions or deletions). The framesearch method produces a series of global or local pairwise alignments between a query nucleotide sequence and a search set of protein sequences, or vice versa. Its ability to evaluate frameshifts offset by an arbitrary number of nucleotides makes the method useful for sequences containing large numbers of indels, which can be very difficult to align with more efficient heuristic methods. In practice, the method requires large amounts of computing power or a system whose architecture is specialized for dynamic programming. The BLAST and EMBOSS suites provide basic tools for creating translated alignments (though some of these approaches take advantage of side-effects of sequence searching capabilities of the tools). More general methods are available from open-source software such as Genewise.

The dynamic programming method is guaranteed to find an optimal alignment given a particular scoring function; however, identifying a good scoring function is often an empirical rather than a theoretical matter. Although dynamic programming is extensible to more than two sequences, it is prohibitively slow for large numbers of sequences or extremely long sequences.

Word Methods

Word methods, also known as *k*-tuple methods, are heuristic methods that are not guaranteed to find an optimal alignment solution, but are significantly more efficient than dynamic programming. These methods are especially useful in large-scale database searches where it is understood that a large proportion of the candidate sequences will have essentially no significant match with the query sequence. Word methods are best known for their implementation in the database search tools FASTA and the BLAST family. Word methods identify a series of short, nonoverlapping subsequences ("words") in the query sequence that are then matched to candidate database sequences. The relative positions of the word in the two sequences being compared are subtracted to obtain an offset; this will indicate a region of alignment if multiple distinct words produce the same offset. Only if this region is detected do these methods apply more sensitive alignment criteria; thus, many unnecessary comparisons with sequences of no appreciable similarity are eliminated.

In the FASTA method, the user defines a value *k* to use as the word length with which to search the database. The method is slower but more sensitive at lower values of *k*, which are also preferred for searches involving a very short query sequence. The BLAST family of search methods provides a number of algorithms optimized for particular types of queries, such as searching for distantly related sequence matches. BLAST was developed to provide a faster alternative to FASTA without sacrificing much accuracy; like FASTA, BLAST uses a word search of length *k*, but evaluates only the most significant word matches, rather than every word match as does FASTA. Most BLAST implementations use a fixed default word length that is optimized for the query and database type, and that is changed only under special circumstances, such as when searching with repetitive or very short query sequences. Implementations can be found via a number of web portals, such as EMBL FASTA and NCBI BLAST.

GLOBAL AND LOCAL ALIGNMENT

Global alignments, which attempt to align every residue in every sequence, are most useful when the sequences in the query set are similar and of roughly equal size. (This does not mean global alignments cannot start and end in gaps). A general global alignment technique is the Needleman–Wunsch algorithm, which is based on dynamic programming. Local alignments are more useful for dissimilar sequences that are suspected to contain regions of similarity or similar sequence motifs within their larger sequence context. The Smith–Waterman algorithm is a general local alignment method based on the same dynamic programming scheme but with additional choices to start and end at any place.

Hybrid methods, known as semi-global or "glocal" (short for global-local) methods, search for the best possible partial alignment of the two sequences (in other words, a combination of one or both starts and one or both ends is stated to be aligned). This can be especially useful when the downstream part of one sequence overlaps with the upstream part of the other sequence. In this case, neither global nor local alignment is entirely appropriate: a global alignment would attempt to force the alignment to extend beyond the region of overlap, while a local alignment might not fully cover the region of overlap. Another case where semi-global alignment is useful is when one sequence is short (for example a gene sequence) and the other is very long (for example a chromosome sequence). In that case, the short sequence should be globally (fully) aligned but only a local (partial) alignment is desired for the long sequence.

Fast expansion of genetic data challenges speed of current DNA sequence alignment algorithms. Essential needs for an efficient and accurate method for DNA variant discovery demand innovative approaches for parallel processing in real time. Optical computing approaches have been suggested as promising alternatives to the current electrical implementations, yet their applicability remains to be tested.

MULTIPLE SEQUENCE ALIGNMENT

A multiple sequence alignment (MSA) is a sequence alignment of three or more biological sequences, generally protein, DNA, or RNA. In many cases, the input set of query sequences are assumed to have an evolutionary relationship by which they share a linkage and are descended from a common ancestor. From the resulting MSA, sequence homology can be inferred and phylogenetic analysis can be conducted to assess the sequences' shared evolutionary origins. Visual depictions of the alignment as in the image at right illustrate mutation events such as point mutations (single amino acid or nucleotide changes) that appear as differing characters in a single alignment column, and insertion or deletion mutations (indels or gaps) that appear as hyphens in one or more of the sequences in the alignment. Multiple sequence alignment is often used to assess sequence conservation of protein domains, tertiary and secondary structures, and even individual amino acids or nucleotides.

Multiple sequence alignment also refers to the process of aligning such a sequence set. Because three or more sequences of biologically relevant length can be difficult and are almost always time-consuming to align by hand, computational algorithms are used to produce and analyze the alignments. MSAs require more sophisticated methodologies than pairwise alignment because they are more computationally complex. Most multiple sequence alignment programs use heuristic methods rather than global optimization because identifying the optimal alignment between more than a few sequences of moderate length is prohibitively computationally expensive.

Algorithm

Multiple sequence alignments can be helpful in many circumstances like detecting historical and familial relations between sequences of proteins or amino acids and determining certain structures or locations on sequences. Therefore it makes sense to construct an algorithm to assist in repetitive calculations of multiple sequence alignments.

Mathematical Definition

Given m sequences $S_i, i = 1, \cdots, m$ similar to the form below:

$$S := \begin{cases} S_1 = (S_{11}, S_{12}, \cdots, S_{1n_1}) \\ S_2 = (S_{21}, S_{22}, \cdots, S_{2n_2}) \\ \vdots \\ S_m = (S_{m1}, S_{m2}, \cdots, S_{mn_m}). \end{cases}$$

A MSA is taken of this set of sequences S by inserting any amount of gaps needed into each of the S_i sequences of S until the modified sequences, S_i', all conform to length $L \geq \max\{x_i \mid i=1,\cdots,r\}$ and no values in the sequences of S of the same column m, consists of only gaps. The mathematical form of an MSA of the above sequence set is shown below:

$$S' := \begin{cases} S_1' = (S_{11}', S_{12}', \cdots, S_{1L}') \\ S_2' = (S_{21}', S_{22}', \cdots, S_{2L}') \\ \vdots \\ S_m' = (S_{r1}', S_{r2}', \cdots, S_{rL}'). \end{cases}$$

To return from each particular sequence S_i' to S_i, remove all gaps.

Graphing Approach

A general approach when calculating multiple sequence alignments is to use graphs to identify all of the different alignments. When finding alignments via graph, a *complete alignment* is created in a weighted graph that contains a set of vertices and a set of edges. Each of the graph edges has a weight based on a certain heuristic that helps to score each *alignment* or subset of the original graph.

Tracing Alignments

When determining the best suited alignments for each MSA, a *trace* is usually generated. A trace is a set of *realized*, or corresponding and aligned, vertices that has a specific weight based on the edges that are selected between corresponding vertices. When choosing traces for a set of sequences it is necessary to choose a trace with a maximum weight to get the best alignment of the sequences.

Application in Code

Dynamic Programming and Computational Complexity

A direct method for producing an MSA uses the dynamic programming technique to identify the globally optimal alignment solution. For proteins, this method usually involves two sets of parameters: a gap penalty and a substitution matrix assigning scores or probabilities to the alignment of each possible pair of amino acids based on the similarity of the amino acids' chemical properties and the evolutionary probability of the mutation. For nucleotide sequences, a similar gap penalty is used, but a much simpler substitution matrix, wherein only identical matches and mismatches are considered, is typical. The scores in the substitution matrix may be either all positive or a mix of positive and negative in the case of a global alignment, but must be both positive and negative, in the case of a local alignment.

For n individual sequences, the naive method requires constructing the n-dimensional equivalent of the matrix formed in standard pairwise sequence alignment. The search space thus increases exponentially with increasing n and is also strongly dependent on sequence length. Expressed with the big O notation commonly used to measure computational complexity, a naïve MSA takes $O(Length^{Nseqs})$ time to produce. To find the global optimum for n sequences this way has been shown to be an NP-complete problem. In 1989, based on Carrillo-Lipman Algorithm, Altschul introduced a practical method that uses pairwise alignments to constrain the n-dimensional search space. In this approach pairwise dynamic programming alignments are performed on each pair of sequences in the query set, and only the space near the n-dimensional intersection of these alignments is searched for the n-way alignment. The MSA program optimizes the sum of all of the pairs of characters at each position in the alignment (the so-called *sum of pair* score) and has been implemented in a software program for constructing multiple sequence alignments.

Alignment Methods

There are various alignment methods used within multiple sequence to maximize scores and correctness of alignments. Each is usually based on a certain heuristic with an insight into the evolutionary process. Most try to replicate evolution to get the most realistic alignment possible to best predict relations between sequences.

Progressive Alignment Construction

The most widely used approach to multiple sequence alignments uses a heuristic search known as progressive technique (also known as the hierarchical or tree method) developed by Da-Fei Feng and Doolittle in 1987. Progressive alignment builds up a final MSA by combining pairwise alignments beginning with the most similar pair and progressing to the most distantly related. All progressive alignment methods require two stages: a first stage in which the relationships between the sequences are represented as a tree, called a *guide tree*, and a second step in which the MSA is built by adding the sequences sequentially to the growing MSA according to the guide tree. The initial *guide tree* is determined by an efficient clustering method such as neighbor-joining or UPGMA, and may use distances based on the number of identical two-letter sub-sequences (as in FASTA rather than a dynamic programming alignment).

Progressive alignments are not guaranteed to be globally optimal. The primary problem is that when errors are made at any stage in growing the MSA, these errors are then propagated through to the final result. Performance is also particularly bad when all of the sequences in the set are rather distantly related. Most modern progressive methods modify their scoring function with a secondary weighting function that assigns scaling factors to individual members of the query set in a nonlinear fashion based on their phylogenetic distance from their nearest neighbors. This corrects for non-random selection of the sequences given to the alignment program.

Progressive alignment methods are efficient enough to implement on a large scale for many (100s to 1000s) sequences. Progressive alignment services are commonly available on publicly accessible web servers so users need not locally install the applications of interest. The most popular progressive alignment method has been the Clustal family, especially the weighted variant ClustalW to which access is provided by a large number of web portals including GenomeNet, EBI, and EMBNet. Different portals or implementations can vary in user interface and make different parameters accessible to the user. ClustalW is used extensively for phylogenetic tree construction, in spite of the author's explicit warnings that unedited alignments should not be used in such studies and as input for protein structure prediction by homology modeling. Current version of Clustal family is ClustalW2. EMBL-EBI announced that CLustalW2 will be expired in August 2015. They recommend Clustal Omega which performs based on seeded guide trees and HMM profile-profile techniques for protein alignments. They offer different MSA tools for progressive DNA alignments. One of them is MAFFT (Multiple Alignment using Fast Fourier Transform).

Another common progressive alignment method called T-Coffee is slower than Clustal and its derivatives but generally produces more accurate alignments for distantly related sequence sets. T-Coffee calculates pairwise alignments by combining the direct alignment of the pair with indirect alignments that aligns each sequence of the pair to a third sequence. It uses the output from Clustal as well as another local alignment program LALIGN, which finds multiple regions of local alignment between two sequences. The resulting alignment and phylogenetic tree are used as a guide to produce new and more accurate weighting factors.

Because progressive methods are heuristics that are not guaranteed to converge to a global optimum, alignment quality can be difficult to evaluate and their true biological significance can be obscure. A semi-progressive method that improves alignment quality and does not use a lossy heuristic while still running in polynomial time has been implemented in the program PSAlign.

Iterative Methods

A set of methods to produce MSAs while reducing the errors inherent in progressive methods are classified as "iterative" because they work similarly to progressive methods but repeatedly realign the initial sequences as well as adding new sequences to the growing MSA. One reason progressive methods are so strongly dependent on a high-quality initial alignment is the fact that these alignments are always incorporated into the final result — that is, once a sequence has been aligned into the MSA, its alignment is not considered further. This approximation improves efficiency at the cost of accuracy. By contrast, iterative methods can return to previously calculated pairwise alignments or sub-MSAs incorporating subsets of the query sequence as a means of optimizing a general objective function such as finding a high-quality alignment score.

A variety of subtly different iteration methods have been implemented and made available in software packages; reviews and comparisons have been useful but generally refrain from choosing a "best" technique. The software package PRRN/PRRP uses a hill-climbing algorithm to optimize its MSA alignment score and iteratively corrects both alignment weights and locally divergent or "gappy" regions of the growing MSA. PRRP performs best when refining an alignment previously constructed by a faster method.

Another iterative program, DIALIGN, takes an unusual approach of focusing narrowly on local alignments between sub-segments or sequence motifs without introducing a gap penalty. The alignment of individual motifs is then achieved with a matrix representation similar to a dot-matrix plot in a pairwise alignment. An alternative method that uses fast local alignments as anchor points or "seeds" for a slower global-alignment procedure is implemented in the CHAOS/DIALIGN suite.

A third popular iteration-based method called MUSCLE (multiple sequence alignment by log-expectation) improves on progressive methods with a more accurate distance measure to assess the relatedness of two sequences. The distance measure is updated between iteration stages (although, in its original form, MUSCLE contained only 2-3 iterations depending on whether refinement was enabled).

Consensus Methods

Consensus methods attempt to find the optimal multiple sequence alignment given multiple different alignments of the same set of sequences. There are two commonly used consensus methods, M-COFFEE and MergeAlign. M-COFFEE uses multiple sequence alignments generated by seven different methods to generate consensus alignments. MergeAlign is capable of generating consensus alignments from any number of input alignments generated using different models of sequence evolution or different methods of multiple sequence alignment. The default option for MergeAlign is to infer a consensus alignment using alignments generated using 91 different models of protein sequence evolution.

Hidden Markov Models

Hidden Markov models are probabilistic models that can assign likelihoods to all possible combinations of gaps, matches, and mismatches to determine the most likely MSA or set of possible MSAs. HMMs can produce a single highest-scoring output but can also generate a family of possible alignments that can then be evaluated for biological significance. HMMs can produce both global and local alignments. Although HMM-based methods have been developed relatively recently, they offer significant improvements in computational speed, especially for sequences that contain overlapping regions.

Typical HMM-based methods work by representing an MSA as a form of directed acyclic graph known as a partial-order graph, which consists of a series of nodes representing possible entries in the columns of an MSA. In this representation a column that is absolutely conserved (that is, that all the sequences in the MSA share a particular character at a particular position) is coded as a single node with as many outgoing connections as there are possible characters in the next column of the alignment. In the terms of a typical hidden Markov model, the observed states are the individual alignment columns and the "hidden" states represent the presumed ancestral sequence from which the sequences in the query set are hypothesized to have descended. An efficient search variant of the dynamic programming method, known as the Viterbi algorithm, is generally used to successively align the growing MSA to the next sequence in the query set to produce a new MSA. This is distinct from progressive alignment methods because the alignment of prior sequences is updated at each new sequence addition. However, like progressive methods, this technique can be influenced by the order in which the sequences in the query set are integrated into the alignment, especially when the sequences are distantly related.

Several software programs are available in which variants of HMM-based methods have been implemented and which are noted for their scalability and efficiency, although properly using an HMM method is more complex than using more common progressive methods. The simplest is POA (Partial-Order Alignment); a similar but more generalized method is implemented in the packages SAM (Sequence Alignment and Modeling System) and HMMER. SAM has been used as a source of alignments for protein structure prediction to participate in the CASP structure prediction experiment and to develop a database of predicted proteins in the yeast species *S. cerevisiae*. HHsearch is a software package for the detection of remotely related protein sequences based on the pairwise comparison of HMMs. A server running HHsearch (HHpred) was by far the fastest of the 10 best automatic structure prediction servers in the CASP7 and CASP8 structure prediction competitions.

Phylogeny-aware Methods

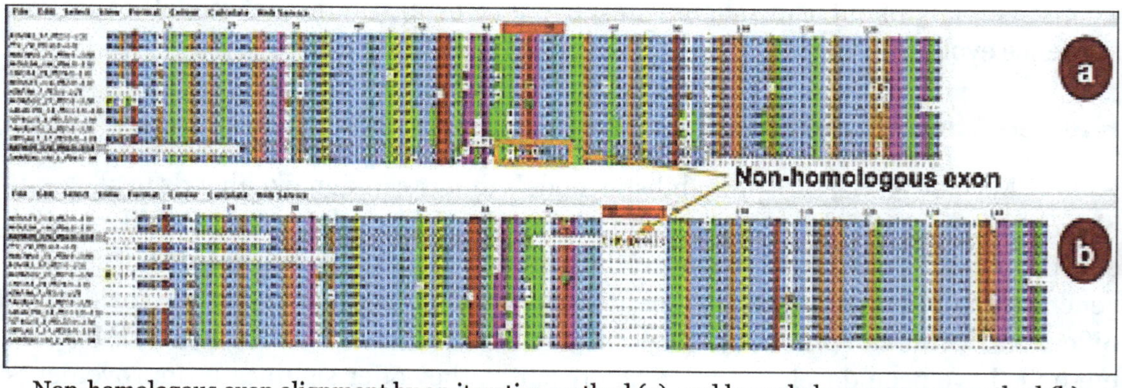

Non-homologous exon alignment by an iterative method (a), and by a phylogeny-aware method (b).

Most multiple sequence alignment methods try to minimize the number of insertions/deletions (gaps) and, as a consequence, produce compact alignments. This causes several problems if the sequences to be aligned contain non-homologous regions, if gaps are informative in a phylogeny analysis. These problems are common in newly produced sequences that are poorly annotated and may contain frame-shifts, wrong domains or non-homologous spliced exons. The first such method was developed in 2005 by Löytynoja and Goldman. The same authors released a software package called *PRANK* in 2008. PRANK improves alignments when insertions are present. Nevertheless, it runs slowly compared to progressive and iterative methods which have been developed for several years.

In 2012, two new phylogeny-aware tools appeared. One is called *PAGAN* that was developed by the same team as PRANK. The other is *ProGraphMSA* developed by Szalkowski. Both software packages were developed independently but share common features, notably the use of graph algorithms to improve the recognition of non-homologous regions, and an improvement in code making these software faster than PRANK.

Motif Finding

Alignment of the seven Drosophila caspases colored by motifs as identified by MEME. When motif positions and sequence alignments are generated independently, they often correlate well but not perfectly, as in this example.

Motif finding, also known as profile analysis, is a method of locating sequence motifs in global MSAs that is both a means of producing a better MSA and a means of producing a scoring matrix for use in searching other sequences for similar motifs. A variety of methods for isolating the motifs have been developed, but all are based on identifying short highly conserved patterns within the larger alignment and constructing a matrix similar to a substitution matrix that reflects the amino acid or nucleotide composition of each position in the putative motif. The alignment can then be refined using these matrices. In standard profile analysis, the matrix includes entries for each possible character as well as entries for gaps. Alternatively, statistical pattern-finding algorithms can identify motifs as a precursor to an MSA rather than as a derivation. In many cases when the query set contains only a small number of sequences or contains only highly related sequences, pseudocounts are added to normalize the distribution reflected in the scoring matrix. In particular, this corrects zero-probability entries in the matrix to values that are small but nonzero.

Blocks analysis is a method of motif finding that restricts motifs to ungapped regions in the alignment. Blocks can be generated from an MSA or they can be extracted from unaligned sequences using a precalculated set of common motifs previously generated from known gene families. Block scoring generally relies on the spacing of high-frequency characters rather than on the calculation of an explicit substitution matrix. The BLOCKS server provides an interactive method to locate such motifs in unaligned sequences.

Statistical pattern-matching has been implemented using both the expectation-maximization algorithm and the Gibbs sampler. One of the most common motif-finding tools, known as MEME, uses expectation maximization and hidden Markov methods to generate motifs that are then used as search tools by its companion MAST in the combined suite MEME/MAST.

Non-coding Multiple Sequence Alignment

Non-coding DNA regions, especially TFBSs, are rather more conserved and not necessarily evolutionarily related, and may have converged from non-common ancestors. Thus, the assumptions used to align protein sequences and DNA coding regions are inherently different from those that hold for TFBS sequences. Although it is meaningful to align DNA coding regions for homologous sequences using mutation operators, alignment of binding site sequences for the same transcription factor cannot rely on evolutionary related mutation operations. Similarly, the evolutionary operator of point mutations can be used to define an edit distance for coding sequences, but this has little meaning for TFBS sequences because any sequence variation has to maintain a certain level of specificity for the binding site to function. This becomes specifically important when trying to align known TFBS sequences to build supervised models to predict unknown locations of the same TFBS. Hence, Multiple Sequence Alignment methods need to adjust the underlying evolutionary hypothesis and the operators used as in the work published incorporating neighbouring base thermodynamic information to align the binding sites searching for the lowest thermodynamic alignment conserving specificity of the binding site, EDNA.

Optimization

Genetic Algorithms and Simulated Annealing

Standard optimization techniques in computer science — both of which were inspired by, but do not directly reproduce, physical processes — have also been used in an attempt to more efficiently produce quality MSAs. One such technique, genetic algorithms, has been used for MSA production in an attempt to broadly simulate the hypothesized evolutionary process that gave rise to the divergence in the query set. The method works by breaking a series of possible MSAs into fragments and repeatedly rearranging those fragments with the introduction of gaps at varying positions. A general objective function is optimized during the simulation, most generally the "sum of pairs" maximization function introduced in dynamic programming-based MSA methods. A technique for protein sequences has been implemented in the software program SAGA (Sequence Alignment by Genetic Algorithm) and its equivalent in RNA is called RAGA.

The technique of simulated annealing, by which an existing MSA produced by another method is refined by a series of rearrangements designed to find better regions of alignment space than the one the input alignment already occupies. Like the genetic algorithm method, simulated annealing maximizes an objective function like the sum-of-pairs function. Simulated annealing uses a metaphorical "temperature factor" that determines the rate at which rearrangements proceed and the likelihood of each rearrangement; typical usage alternates periods of high rearrangement rates with relatively low likelihood (to explore more distant regions of alignment space) with periods of lower rates and higher likelihoods to more thoroughly explore local minima near the newly "colonized" regions. This approach has been implemented in the program MSASA (Multiple Sequence Alignment by Simulated Annealing).

Alignment Visualization and Quality Control

The necessary use of heuristics for multiple alignment means that for an arbitrary set of proteins, there is always a good chance that an alignment will contain errors. For example, an evaluation of several leading alignment programs using the BAliBase benchmark found that at least 24% of all pairs of aligned amino acids were incorrectly aligned. These errors can arise because of unique insertions into one or more regions of sequences, or through some more complex evolutionary process leading to proteins that do not align easily by sequence alone. As the number of sequence and their divergence increases many more errors will be made simply because of the heuristic nature of MSA algorithms. Multiple sequence alignment viewers enable alignments to be visually reviewed, often by inspecting the quality of alignment for annotated functional sites on two or more sequences. Many also enable the alignment to be edited to correct these (usually minor) errors, in order to obtain an optimal 'curated' alignment suitable for use in phylogenetic analysis or comparative modeling.

However, as the number of sequences increases and especially in genome-wide studies that involve many MSAs it is impossible to manually curate all alignments. Furthermore, manual curation is subjective. And finally, even the best expert cannot confidently align the more ambiguous cases of highly diverged sequences. In such cases it is common practice to use automatic procedures to exclude unreliably aligned regions from the MSA. For the purpose of phylogeny reconstruction the Gblocks program is widely used to remove alignment blocks suspect of low quality, according to various cutoffs on the number of gapped sequences in alignment columns. However, these criteria may excessively filter out regions with insertion/deletion events that may still be aligned reliably, and these regions might be desirable for other purposes such as detection of positive selection. A few alignment algorithms output site-specific scores that allow the selection of high-confidence regions. Such a service was first offered by the SOAP program, which tests the robustness of each column to perturbation in the parameters of the popular alignment program CLUSTALW. The T-Coffee program uses a library of alignments in the construction of the final MSA, and its output MSA is colored according to confidence scores that reflect the agreement between different alignments in the library regarding each aligned residue. Its extension, TCS : (Transitive Consistency Score), uses T-Coffee libraries of pairwise alignments to evaluate any third party MSA. Pairwise projections can be produced using fast or slow methods, thus allowing a trade-off between speed and accuracy. Another alignment program that can output an MSA with confidence scores is FSA, which uses a statistical model that allows calculation of the uncertainty in the alignment. The HoT (Heads-Or-Tails) score can be used as a measure of site-specific alignment uncertainty due to the existence of multiple co-optimal solutions. The GUIDANCE program calculates a similar site-specific confidence measure based on the robustness of the alignment to uncertainty in the guide tree that is used in progressive alignment programs. An alternative, more statistically justified approach to assess alignment uncertainty is the use of probabilistic evolutionary models for joint estimation of phylogeny and alignment. A Bayesian approach allows calculation of posterior probabilities of estimated phylogeny and alignment, which is a measure of the confidence in these estimates. In this case, a posterior probability can be calculated for each site in the alignment. Such an approach was implemented in the program BAli-Phy.

There free available programs for visualization of multiple sequence alignments: JalView, UGENE.

Phylogenetic Use

Multiple sequence alignments can be used to create a phylogenetic tree. This is made possible by two reasons. The first is because functional domains that are known in annotated sequences can be used for alignment in non-annotated sequences. The other is that conserved regions known to be functionally important can be found. This makes it possible for multiple sequence alignments to be used to analyze and find evolutionary relationships through homology between sequences. Point mutations and insertion or deletion events (called indels) can be detected.

Multiple sequence alignments can also be used to identify functionally important sites, such as binding sites, active sites, or sites corresponding to other key functions, by locating conserved domains. When looking at multiple sequence alignments, it is useful to consider different aspects of the sequences when comparing sequences. These aspects include identity, similarity, and homology. Identity means that the sequences have identical residues at their respective positions. On the other hand, similarity has to do with the sequences being compared having similar residues quantitatively. For example, in terms of nucleotide sequences, pyrimidines are considered similar to each other, as are purines. Similarity ultimately leads to homology, in that the more similar sequences are, the closer they are to being homologous. This similarity in sequences can then go on to help find common ancestry.

Application of Multiple Sequence Alignment

The main applications of Multiple Sequence Alignment are:

- Structure Prediction: A Multiple Sequence Alignment can give you the almost perfect protein or RNA secondary structure; sometimes it helps even with the 3D structure.
- Protein Family: A Multiple Sequence Alignment can help you to decide that your protein is a member of a known protein family or not.
- Pattern Identification: By looking at conserved regions or sites, you can identify which region is responsible for a functional site.
- Domain Identification: By looking at file provided by a Multiple Sequence Alignment, you can extract profiles to use them against databases.
- DNA Regulatory Elements: You can use Multiple Sequence Alignments to locate DNA regulatory elements such as binding sites, etc.
- Phylogenetic Analysis: By carefully picking related sequences you can reconstruct a tree using sequences that u have used in the Multiple Sequence Alignment.

DNA SEQUENCING

Sequencing is the operation of determining the precise order of nucleotides of a given DNA molecule. It is used to determine the order of the four bases adenine (A), guanine (G), cytosine (C) and thymine (T), in a strand of DNA.

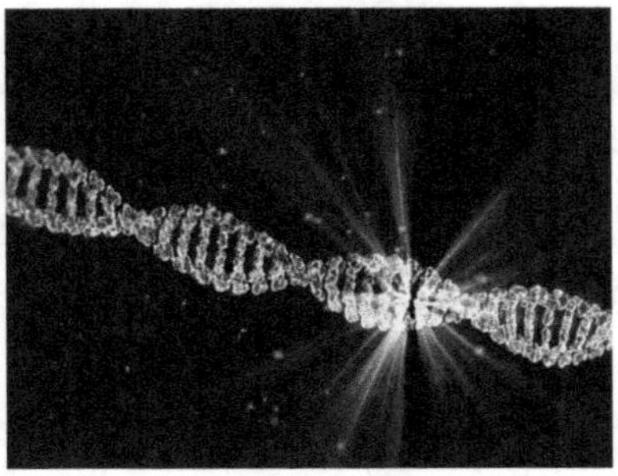

DNA molecule.

DNA sequencing is used to determine the sequence of individual genes, full chromosomes or entire genomes of an organism. DNA sequencing has also become the most efficient way to sequence RNA or proteins.

Whole Genome Sequencing and Sequence Assembly

A DNA sequencing reaction produces a sequence that is several hundred bases long.

Gene sequences are typically thousands of bases long. The largest known gene is the one associated with Duchenne muscular dystrophy. It is approximately 2.4 million bases in length. In order to study one whole gene, scientists use a simple strategy known as shotgun sequencing. The long DNA sequence is assembled from a series of shorter overlapping sequences.

Shotgun Sequencing

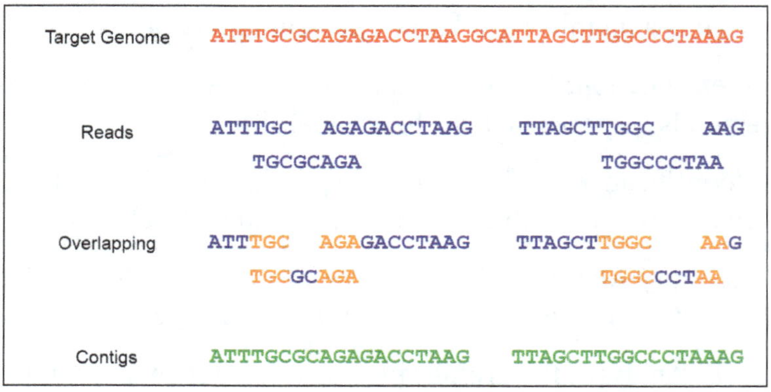

Shotgun sequencing.

Special machines, known as sequencing machines are used to extract short random DNA sequences from a particular genome we wish to determine (target genome). Current DNA sequencing technologies cannot read one whole genome at once. It reads small pieces of between 20 and 30000 bases, depending on the technology used. These short pieces are called reads. Special software are used to assemble these reads according to how they overlap, in order to generate continuous strings called contigs. These contigs can be the whole target genome itself, or parts of the genome.

The process of aligning and merging fragments from a longer DNA sequence, in order to reconstruct the original sequence is known as Sequence Assembly.

In order to obtain the whole genome sequence, we may need to generate more and more random reads, until the contigs match to the target genome.

Sequence Assembly Problem

The sequence assembly problem can be described as follows.

Given a set of sequences, find the minimal length string containing all members of the set as substrings.

This problem is further complicated due to the existence of repetitive sequences in the genome as well as substitutions or mutations withing them. The sequence assembly problem can be compared to a real life scenario as follows.

It sounds very confusing and quite impossible to be carried out. This problem is known to be NP Complete. NP complete problems are problems whose status is unknown. No polynomial time algorithm has yet been discovered for any NP complete problem, nor has anybody yet been able to prove that no polynomial-time algorithm exists for any of them. However, there are greedy algorithms to solve the sequence assembly problem, where experiments have proven to perform fairly well in practice. A common method used to solve the sequence assembly problem and perform sequence data analysis is sequence alignment.

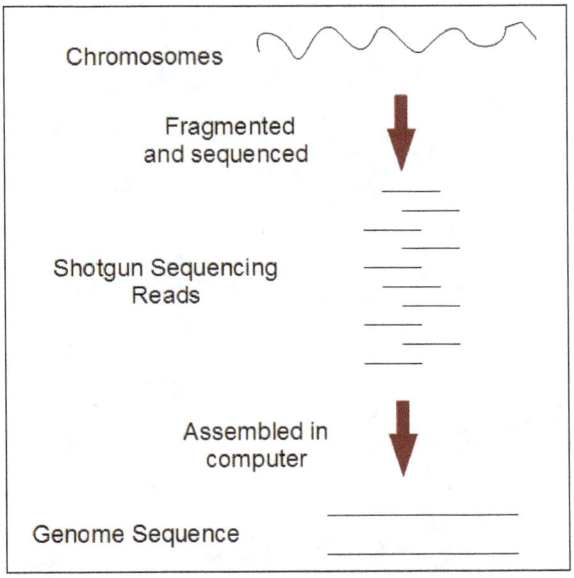

Sequence Alignment

Sequence alignment is a method of arranging sequences of DNA, RNA, or protein to identify regions of similarity. The similarity being identified may be a result of functional, structural, or evolutionary relationships between the sequences. If we compare two sequences, it is known as pairwise sequence alignment. If we compare more than two sequences, it is known as multiple sequence alignment.

Next-generation Sequencing

Next-generation sequencing (NGS), also known as high-throughput sequencing, is the collective term used to describe many different modern sequencing technologies such as:

- Illumina (Solexa) sequencing.
- Roche 454 sequencing.
- Ion torrent Proton/PGM sequencing.
- SOLiD sequencing.

HUMAN GENOME PROJECT

Human Genome Project (HGP) is an international collaboration that successfully determined, stored, and rendered publicly available the sequences of almost all the genetic content of the chromosomes of the human organism, otherwise known as the human genome.

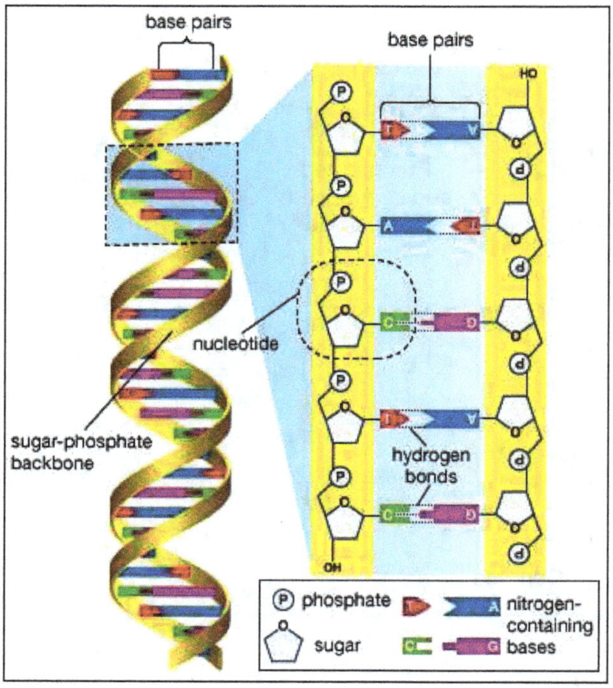

The human genome is made up of approximately three billion base pairs of deoxyribonucleic acid (DNA). The bases of DNA are adenine (A), thymine (T), guanine (G), and cytosine (C).

The Human Genome Project (HGP), which operated from 1990 to 2003, provided researchers with basic information about the sequences of the three billion chemical base pairs (i.e., adenine [A], thymine [T], guanine [G], and cytosine [C]) that make up human genomic DNA (deoxyribonucleic acid). The HGP was further intended to improve the technologies needed to interpret and analyze genomic sequences, to identify all the genes encoded in human DNA, and to address the ethical, legal, and social implications that might arise from defining the entire human genomic sequence.

Science behind the HGP

To appreciate the magnitude, challenge, and implications of the HGP, it is important first to consider the foundation of science upon which it was based—the fields of classical, molecular, and human genetics. Classical genetics is considered to have begun in the mid-1800s with the work of Austrian botanist, teacher, and Augustinian prelate

Gregor Mendel, who defined the basic laws of genetics in his studies of the garden pea (Pisum sativum). Mendel succeeded in explaining that, for any given gene, offspring inherit from each parent one form, or allele, of a gene. In addition, the allele that an offspring inherits from a parent for one gene is independent of the allele inherited from that parent for another gene.

Mendel's basic laws of genetics were expanded upon in the early 20th century when molecular geneticists began conducting research using model organisms such as Drosophila melanogaster (also called the vinegar fly or fruit fly) that provided a more comprehensive view of the complexities of genetic transmission. For example, molecular genetics studies demonstrated that two alleles can be codominant (characteristics of both alleles of a gene are expressed) and that not all traits are defined by single genes; in fact, many traits reflect the combined influences of numerous genes. The field of molecular genetics emerged from the realization that DNA and RNA (ribonucleic acid) constitute the genetic material in all living things. In physical terms, a gene is a discrete stretch of nucleotides within a DNA molecule, with each nucleotide containing an A, G, T, or C base unit. It is the specific sequence of these bases that encodes the information contained in the gene and that is ultimately translated into a final product, a molecule of protein or in some cases a molecule of RNA. The protein or RNA product may have a structural role or a regulatory role, or it may serve as an enzyme to promote the formation or metabolism of other molecules, including carbohydrates and lipids. All these molecules work in concert to maintain the processes required for life.

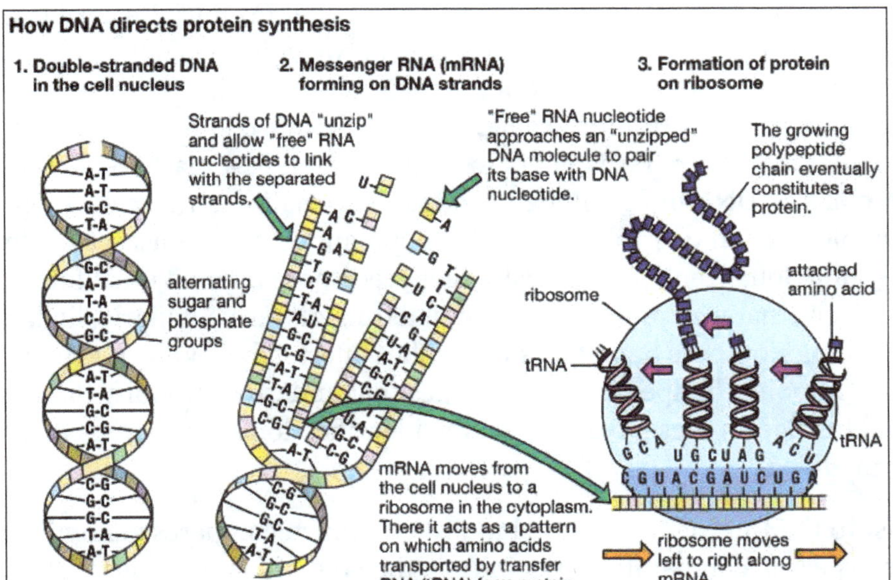

Molecular genetics emerged from the realization that DNA and RNA constitute the genetic material of all living organisms. (1) DNA, located in the cell nucleus, is made up of nucleotides that contain the bases adenine (A), thymine (T), guanine (G), and cytosine (C). (2) RNA, which contains uracil (U) instead of thymine, transports the genetic code to protein-synthesizing sites in the cell. (3) Messenger RNA (mRNA) then carries the genetic information to ribosomes in the cell cytoplasm that translate the genetic information into molecules of protein.

Studies in molecular genetics led to studies in human genetics and the consideration of the ways in which traits in humans are inherited. For example, most traits in humans and other species result from a combination of genetic and environmental influences. In addition, some genes, such as those encoded at neighbouring spots on a single chromosome, tend to be inherited together, rather than independently, whereas other genes, namely those encoded on the mitochondrial genome, are inherited only from the mother, and yet other genes, encoded on the Y chromosome, are passed only from fathers to sons. Using data from the HGP, scientists have estimated that the human genome contains anywhere from 20,000 to 25,000 genes.

Advances based on the HGP

Advances in genetics and genomics continue to emerge. Two important advances include the International HapMap Project and the initiation of large-scale comparative genomics studies, both of which have been made possible by the availability of databases of genomic sequences of humans, as well as the availability of databases of genomic sequences of a multitude of other species.

The International HapMap Project is a collaborative effort between Japan, the United Kingdom, Canada, China, Nigeria, and the United States in which the goal is to identify and catalog genetic similarities and differences between individuals representing four major human populations derived from the continents of Africa, Europe, and Asia. The identification of genetic variations called polymorphisms that exist in DNA sequences among populations allows researchers to define haplotypes, markers that distinguish specific regions of DNA in the human genome. Association studies of the prevalence of these haplotypes in control and patient populations can be used to help identify potentially functional genetic differences that predispose an individual toward disease or, alternatively, that may protect an individual from disease. Similarly, linkage studies of the inheritance of these haplotypes in families affected by a known genetic trait can also help to pinpoint the specific gene or genes that underlie or modify that trait. Association and linkage studies have enabled the identification of numerous disease genes and their modifiers.

In contrast to the International HapMap Project, which compares genomic sequences within one species, comparative genomics is the study of similarities and differences between different species. In recent years a staggering number of full or almost full genome sequences from different species have been determined and deposited in public databases such as NIH's Entrez Genome database. By comparing these sequences, often using a software tool called BLAST (Basic Local Alignment Search Tool), researchers are able to identify degrees of similarity and divergence between the genes and genomes of related or disparate species. The results of these studies have illuminated the evolution of species and of genomes. Such studies have also helped to draw attention to highly conserved regions of noncoding sequences of DNA that were originally thought to be nonfunctional because they do not contain base sequences that are translated into protein. However, some noncoding regions of DNA have been highly conserved and may play key roles in human evolution.

Impacts of the HGP on Medicine

The public availability of a complete human genome sequence represented a defining moment for both the biomedical community and for society. In the years since completion of the HGP, the human genome database, together with other publicly available resources such as the HapMap database, has enabled the identification of a variety of genes that are associated with disease. This, in turn, has enabled more objective and accurate diagnoses, in some cases even before the onset of overt clinical symptoms. Association and linkage studies have identified additional genetic influences that modify the development or outcome for both rare and common diseases. The recognition that human genomes may influence everything from disease risk to physiological response to medications has led to the emergence of the concept of personalized medicine–the idea that knowledge of a patient's entire genome sequence will give health care providers the ability to deliver the most appropriate and effective care for that patient. Indeed, continuing advances in DNA sequencing technology promise to lower the cost of sequencing an individual's entire genome to that of other, relatively inexpensive, diagnostic tests.

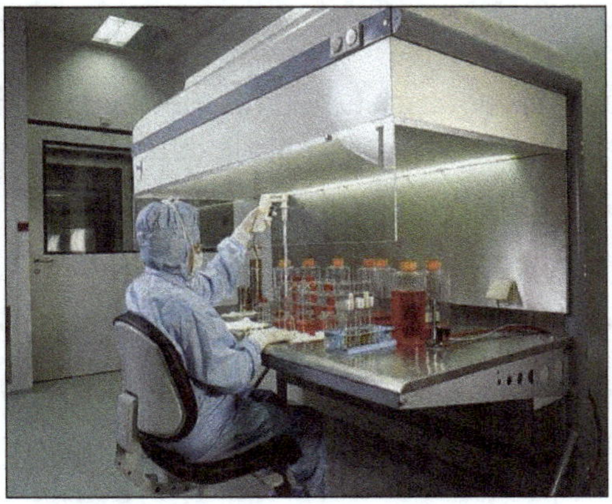

Researchers working in laboratories can determine the genetic composition of an individual's genome.

SEQUENCE ANALYSIS

In bioinformatics, sequence analysis is the process of subjecting a DNA, RNA or peptide sequence to any of a wide range of analytical methods to understand its features, function, structure, or evolution. Methodologies used include sequence alignment, searches against biological databases, and others. Since the development of methods of high-throughput production of gene and protein sequences, the rate of addition of new sequences to the databases increased exponentially. Such a collection of sequences does not, by itself, increase the scientist's understanding of the biology of organisms. However, comparing these new sequences to those with known functions is a key way of understanding the biology of an organism from which the new sequence comes. Thus, sequence analysis can be used to assign function to genes and proteins by the study of the similarities between the compared sequences. Nowadays, there are many tools and techniques that provide the sequence comparisons (sequence alignment) and analyze the alignment product to understand its biology.

Sequence analysis in molecular biology includes a very wide range of relevant topics:

- The comparison of sequences in order to find similarity, often to infer if they are related (homologous).
- Identification of intrinsic features of the sequence such as active sites, post translational modification sites, gene-structures, reading frames, distributions of introns and exons and regulatory elements.
- Identification of sequence differences and variations such as point mutations and single nucleotide polymorphism (SNP) in order to get the genetic marker.
- Revealing the evolution and genetic diversity of sequences and organisms.
- Identification of molecular structure from sequence alone.

In chemistry, sequence analysis comprises techniques used to determine the sequence of a polymer formed of several monomers. In molecular biology and genetics, the same process is called simply "sequencing".

In marketing, sequence analysis is often used in analytical customer relationship management applications, such as NPTB models (Next Product to Buy).

In sociology, sequence methods are increasingly used to study life-course and career trajectories, patterns of organizational and national development, conversation and interaction structure, and the problem of work/family synchrony. This body of research has given rise to the emerging subfield of social sequence analysis.

Sequence Alignment

There are millions of protein and nucleotide sequences known. These sequences fall into many groups of related sequences known as protein families or gene families. Relationships between these sequences are usually discovered by aligning them together and assigning this alignment a score. There are two main types of sequence alignment. Pair-wise sequence alignment only compares two sequences at a time and multiple sequence alignment compares many sequences. Two important algorithms for aligning pairs of sequences are the Needleman-Wunsch algorithm and the Smith-Waterman algorithm. Popular tools for sequence alignment include:

- Pair-wise alignment: BLAST, Dot plots.
- Multiple alignment: ClustalW, PROBCONS, MUSCLE, MAFFT, and T-Coffee.

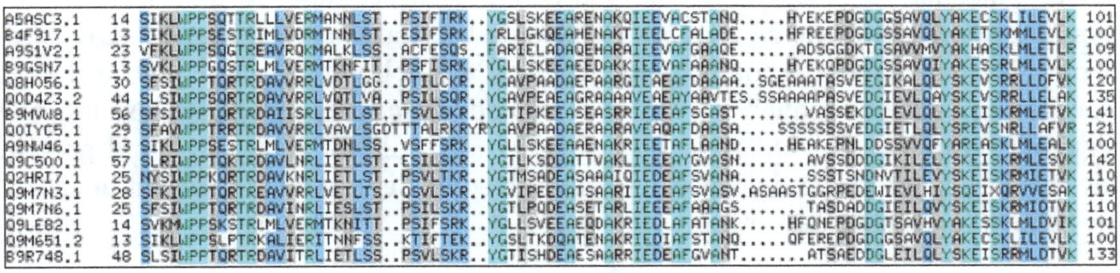

Example multiple sequence alignment.

A common use for pairwise sequence alignment is to take a sequence of interest and compare it to all known sequences in a database to identify homologous sequences. In general, the matches in the database are ordered to show the most closely related sequences first, followed by sequences with diminishing similarity. These matches are usually reported with a measure of statistical significance such as an Expectation value.

Sequence Assembly

Sequence assembly refers to the reconstruction of a DNA sequence by aligning and merging small DNA fragments. It is an integral part of modern DNA sequencing. Since presently-available DNA sequencing technologies are ill-suited for reading long sequences, large pieces of DNA (such as genomes) are often sequenced by: (1) Cutting the DNA into small pieces, (2) Reading the small fragments, and (3) Reconstituting the original DNA by merging the information on various fragments.

Recently, sequencing multiple species at one time is one of the top research objectives. Metagenomics is the study of microbial communities directly obtained from the environment. Different from cultured microorganisms from the lab, the wild sample usually contains dozens, sometimes even thousands of types of microorganisms from their original habitats. Recovering the original genomes can prove to be very challenging.

Gene Prediction

Gene prediction or gene finding refers to the process of identifying the regions of genomic DNA that encode genes. This includes protein-coding genes as well as RNA genes, but may also include the prediction of other functional elements such as regulatory regions. Gene finding is one of the first and most important steps in understanding the genome of a species once it has been sequenced. In general, the prediction of bacterial genes is significantly simpler and more accurate than the prediction of genes in eukaryotic species that usually have complex intron/exon patterns. Identifying genes in long sequences remains a problem, especially when the number of genes is unknown. Hidden markov models can be part of the solution. Machine learning has played a significant role in predicting the sequence of transcription factors. Traditional sequencing analysis focused on the statistical parameters of the nucleotide sequence itself. Another method is to identify homologous sequences based on other known gene sequences. The two methods described here are focused on the sequence. However, the shape feature of these molecules such as DNA and protein have also been studied and proposed to have an equivalent, if not higher, influence on the behaviors of these molecules.

Protein Structure Prediction

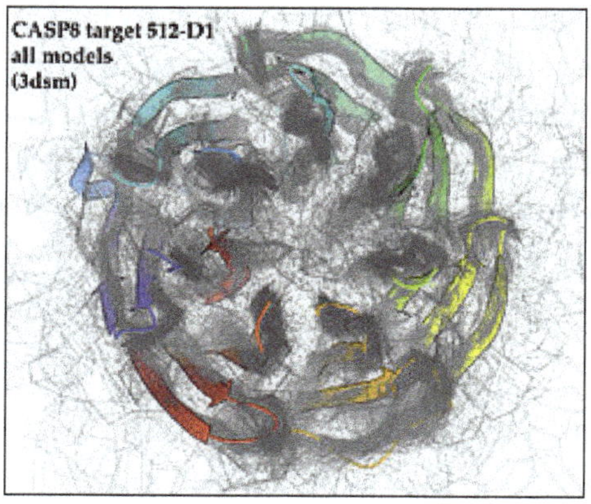

Target protein structure (3dsm, shown in ribbons), with Calpha backbones (in gray) of 354 predicted models for it submitted in the CASP8 structure-prediction experiment.

The 3D structures of molecules are of great importance to their functions in nature. Since structural prediction of large molecules at an atomic level is a largely intractable problem, some biologists introduced ways to predict 3D structure at a primary sequence level. This includes the biochemical or statistical analysis of amino acid residues in local regions and structural the inference from homologs (or other potentially related proteins) with known 3D structures.

There have been a large number of diverse approaches to solve the structure prediction problem. In order to determine which methods were most effective, a structure prediction competition was founded called CASP (Critical Assessment of Structure Prediction).

Methodology

The tasks that lie in the space of sequence analysis are often non-trivial to resolve and require the use of relatively complex approaches. Of the many types of methods used in practice, the most popular include:

- DNA patterns.
- Dynamic Programming.
- Artificial Neural Network.
- Hidden Markov Model.
- Support Vector Machine.
- Clustering.
- Bayesian Network.
- Regression Analysis.
- Sequence Mining.
- Alignment-free Sequence Analysis.

SHOTGUN SEQUENCING

In genetics, shotgun sequencing is a method used for sequencing random DNA strands. It is named by analogy with the rapidly expanding, quasi-random firing pattern of a shotgun.

The chain termination method of DNA sequencing ("Sanger sequencing") can only be used for short DNA strands of 100 to 1000 base pairs. Due to this size limit, longer sequences are subdivided into smaller fragments that can be sequenced separately, and these sequences are assembled to give the overall sequence.

There are two principal methods for this fragmentation and sequencing process. Primer walking (or "chromosome walking") progresses through the entire strand piece by piece, whereas shotgun sequencing is a faster but more complex process that uses random fragments.

In shotgun sequencing, DNA is broken up randomly into numerous small segments, which are sequenced using the chain termination method to obtain *reads*. Multiple overlapping reads for the target DNA are obtained by performing several rounds of this fragmentation and sequencing. Computer programs then use the overlapping ends of different reads to assemble them into a continuous sequence.

Shotgun sequencing was one of the precursor technologies that was responsible for enabling full genome sequencing.

Example:

For example, consider the following two rounds of shotgun reads:

Strand	Sequence
Original	AGCATGCTGCAGTCATGCTTAGGCTA
First shotgun sequence	AGCATGCTGCAGTCATGCT------- -------------------TAGGCTA
Second shotgun sequence	AGCATG-------------------- ------CTGCAGTCATGCTTAGGCTA
Reconstruction	AGCATGCTGCAGTCATGCTTAGGCTA

In this extremely simplified example, none of the reads cover the full length of the original sequence, but the four reads can be assembled into the original sequence using the overlap of their ends to align and order them. In reality, this process uses enormous amounts of information that are rife with ambiguities and sequencing errors. Assembly of complex genomes is additionally complicated by the great abundance of repetitive sequences, meaning similar short reads could come from completely different parts of the sequence.

Many overlapping reads for each segment of the original DNA are necessary to overcome these difficulties and accurately assemble the sequence. For example, to complete the Human Genome Project, most of the human genome was sequenced at 12X or greater *coverage*; that is, each base in the final sequence was present on average in 12 different reads. Even so, current methods have failed to isolate or assemble reliable sequence for approximately 1% of the (euchromatic) human genome, as of 2004.

Whole Genome Shotgun Sequencing

Paired-end Sequencing

Broader application benefited from pairwise end sequencing, known colloquially as *double-barrel shotgun sequencing*. As sequencing projects began to take on longer and more complicated DNA sequences, multiple groups began to realize that useful information could be obtained by sequencing both ends of a fragment of DNA. Although sequencing both ends of the same fragment and keeping track of the paired data was more cumbersome than sequencing a single end of two distinct fragments, the knowledge that the two sequences were oriented in opposite directions and were about the length of a fragment apart from each other was valuable in reconstructing the sequence of the original target fragment.

The first published description of the use of paired ends was in 1990 as part of the sequencing of the human HGPRT locus, although the use of paired ends was limited to closing gaps after the application of a traditional shotgun sequencing approach. The first theoretical description of a pure pairwise end sequencing strategy, assuming fragments of constant length, was in 1991. At the time, there was community consensus that the optimal fragment length for pairwise end sequencing would be three times the sequence read length. In 1995 Roach et al. introduced the innovation of using fragments of varying sizes, and demonstrated that a pure pairwise end-sequencing strategy would be possible on large targets. The strategy was subsequently adopted by The Institute for Genomic Research (TIGR) to sequence the genome of the bacterium *Haemophilus influenzae* in 1995, and then by Celera Genomics to sequence the *Drosophila melanogaster* (fruit fly) genome in 2000, and subsequently the human genome.

Approach

To apply the strategy, a high-molecular-weight DNA strand is sheared into random fragments, size-selected (usually 2, 10, 50, and 150 kb), and cloned into an appropriate vector. The clones are then sequenced from both ends using the chain termination method yielding two short sequences. Each sequence is called an *end-read* or *read* and two reads from the same clone are referred to as *mate pairs*. Since the chain termination method usually can only produce reads between 500 and 1000 bases long, in all but the smallest clones, mate pairs will rarely overlap.

Assembly

The original sequence is reconstructed from the reads using sequence assembly software. First, overlapping reads are collected into longer composite sequences known as *contigs*. Contigs can be linked together into *scaffolds* by following connections between mate pairs. The distance between contigs can be inferred from the mate pair positions if the average fragment length of the library is known and has a narrow window of deviation. Depending on the size of the gap between contigs, different techniques can be used to find the sequence in the gaps. If the gap is small (5-20kb) then the use of polymerase chain reaction (PCR) to amplify the region is required, followed by sequencing. If the gap is large (>20kb) then the large fragment is cloned in special vectors such as bacterial artificial chromosomes (BAC) followed by sequencing of the vector.

Pros and Cons

Proponents of this approach argue that it is possible to sequence the whole genome at once using large arrays of sequencers, which makes the whole process much more efficient than more traditional approaches. Detractors argue that although the technique quickly sequences large regions of DNA, its ability to correctly link these regions is suspect, particularly for genomes with repeating regions. As sequence assembly programs become more sophisticated and computing power becomes cheaper, it may be possible to overcome this limitation.

Coverage

Coverage (read depth or depth) is the average number of reads representing a given nucleotide in the reconstructed sequence. It can be calculated from the length of the original genome (*G*), the number of reads(*N*), and the average read length(*L*) as $N \times L / G$. For example, a hypothetical genome with 2,000 base pairs reconstructed from 8 reads with an average length of 500 nucleotides will have 2x redundancy. This parameter also enables one to estimate other quantities, such as the percentage of the genome covered by reads (sometimes also called coverage). A high coverage in shotgun sequencing is desired because it can overcome errors in base calling and assembly. The subject of DNA sequencing theory addresses the relationships of such quantities.

Sometimes a distinction is made between *sequence coverage* and *physical coverage*. Sequence coverage is the average number of times a base is read. Physical coverage is the average number of times a base is read or spanned by mate paired reads.

Hierarchical Shotgun Sequencing

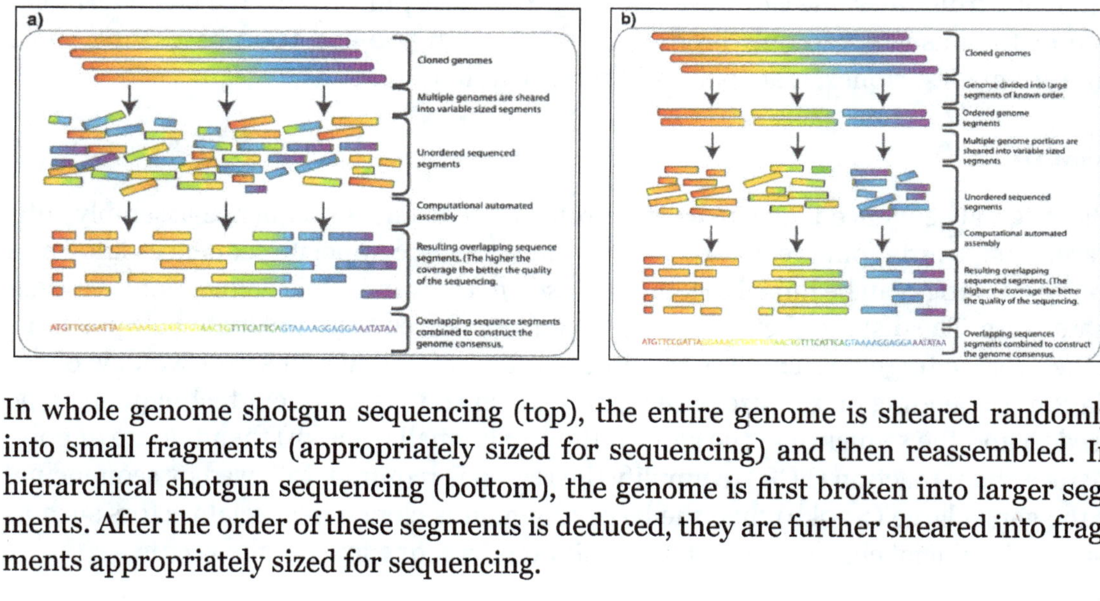

In whole genome shotgun sequencing (top), the entire genome is sheared randomly into small fragments (appropriately sized for sequencing) and then reassembled. In hierarchical shotgun sequencing (bottom), the genome is first broken into larger segments. After the order of these segments is deduced, they are further sheared into fragments appropriately sized for sequencing.

Although shotgun sequencing can in theory be applied to a genome of any size, its direct application to the sequencing of large genomes (for instance, the human genome) was limited until the late 1990s, when technological advances made practical the handling of the vast quantities of complex data involved in the process. Historically, full-genome shotgun sequencing was believed to be limited by both the sheer size of large genomes and by the complexity added by the high percentage of repetitive DNA (greater than 50% for the human genome) present in large genomes. It was not widely accepted that a full-genome shotgun sequence of a large genome would provide reliable data. For these reasons, other strategies that lowered the computational load of sequence assembly had to be utilized before shotgun sequencing was performed. In hierarchical sequencing, also known as top-down sequencing, a low-resolution physical map of the genome is made prior to actual sequencing. From this map, a minimal number of fragments that cover the entire chromosome are selected for sequencing.

In this way, the minimum amount of high-throughput sequencing and assembly is required.

The amplified genome is first sheared into larger pieces (50-200kb) and cloned into a bacterial host using BACs or P1-derived artificial chromosomes (PAC). Because multiple genome copies have been sheared at random, the fragments contained in these clones have different ends, and with enough coverage finding a scaffold of BAC contigs that covers the entire genome is theoretically possible. This scaffold is called a tiling path.

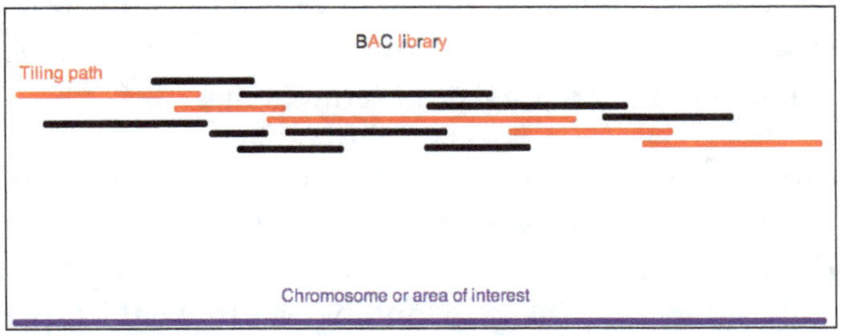

A BAC contig that covers the entire genomic area of interest makes up the tiling path.

Once a tiling path has been found, the BACs that form this path are sheared at random into smaller fragments and can be sequenced using the shotgun method on a smaller scale.

Although the full sequences of the BAC contigs is not known, their orientations relative to one another are known. There are several methods for deducing this order and selecting the BACs that make up a tiling path. The general strategy involves identifying the positions of the clones relative to one another and then selecting the least number of clones required to form a contiguous scaffold that covers the entire area of interest. The order of the clones is deduced by determining the way in which they overlap. Overlapping clones can be identified in several ways. A small radioactively or chemically labeled probe containing a sequence-tagged site (STS) can be hybridized onto a microarray upon which the clones are printed. In this way, all the clones that contain a particular sequence in the genome are identified. The end of one of these clones can then be sequenced to yield a new probe and the process repeated in a method called chromosome walking.

Alternatively, the BAC library can be restriction-digested. Two clones that have several fragment sizes in common are inferred to overlap because they contain multiple similarly spaced restriction sites in common. This method of genomic mapping is called restriction fingerprinting because it identifies a set of restriction sites contained in each clone. Once the overlap between the clones has been found and their order relative to the genome known, a scaffold of a minimal subset of these contigs that covers the entire genome is shotgun-sequenced.

Because it involves first creating a low-resolution map of the genome, hierarchical shotgun sequencing is slower than whole-genome shotgun sequencing, but relies less heavily on computer algorithms than whole-genome shotgun sequencing. The process of extensive BAC library creation and tiling path selection, however, make hierarchical shotgun sequencing slow and labor-intensive. Now that the technology is available and the reliability of the data demonstrated, the speed and cost efficiency of whole-genome shotgun sequencing has made it the primary method for genome sequencing.

Newer Sequencing Technologies

The classical shotgun sequencing was based on the Sanger sequencing method: this was the most advanced technique for sequencing genomes from about 1995–2005. The shotgun strategy is still applied today, however using other sequencing technologies, such as short-read sequencing and long-read sequencing.

Short-read or "next-gen" sequencing produces shorter reads (anywhere from 25–500bp) but many hundreds of thousands or millions of reads in a relatively short time (on the order of a day). This results in high coverage, but the assembly process is much more computationally intensive. These technologies are vastly superior to Sanger sequencing due to the high volume of data and the relatively short time it takes to sequence a whole genome.

Metagenomic Shotgun Sequencing

Having reads of 400-500 base pairs length is sufficient to determine the species/strain of the organism where the DNA comes from, provided its genome is already known, by using for example a k-mer based taxonomic classifier software. With millions of reads from next generation sequencing of an environmental sample, it is possible to get a complete overview of any complex microbiome with thousands of species, like the gut flora. Advantages over 16S rRNA amplicon sequencing are: not being limited to bacteria; strain-level classification where amplicon sequencing only gets the genus; and the possibility to extract whole genes and specify their function as part of the metagenome. The sensitivity of metagenomic sequencing makes it an attractive choice for clinical use. It however emphasizes the problem of contamination of the sample or the sequencing pipeline.

NEXT-GENERATION SEQUENCING

Massive parallel sequencing or massively parallel sequencing is any of several high-throughput approaches to DNA sequencing using the concept of massively parallel processing; it is also called next-generation sequencing (NGS) or second-generation sequencing. Some of these technologies emerged in 1994-1998 and have been commercially available since 2005. These technologies use miniaturized and parallelized platforms for sequencing of 1 million to 43 billion short reads (50-400 bases each) per instrument run.

Many NGS platforms differ in engineering configurations and sequencing chemistry. They share the technical paradigm of massive parallel sequencing via spatially separated, clonally amplified DNA templates or single DNA molecules in a flow cell. This design is very different from that of Sanger sequencing—also known as capillary sequencing or first-generation sequencing—that is based on electrophoretic separation of chain-termination products produced in individual sequencing reactions.

NGS Platforms

DNA sequencing with commercially available NGS platforms is generally conducted with the following steps. First, DNA sequencing libraries are generated by clonal amplification by PCR in vitro. Second, the DNA is sequenced by synthesis, such that the DNA sequence is determined by the addition of nucleotides to the complementary strand rather than through chain-termination chemistry. Third, the spatially segregated, amplified DNA templates are sequenced simultaneously in a massively parallel fashion without the requirement for a physical separation step. While these steps are followed in most NGS platforms, each utilizes a different strategy.

NGS parallelization of the sequencing reactions generates hundreds of megabases to gigabases of nucleotide sequence reads in a single instrument run. This has enabled a drastic increase in available sequence data and fundamentally changed genome sequencing approaches in the biomedical sciences. Newly emerging NGS technologies and instruments have further contributed to a significant decrease in the cost of sequencing nearing the mark of $1000 per genome sequencing.

As the pace of NGS technologies is advancing rapidly, technical specifications and pricing are in flux.

NGS Platforms					
Platform	Template Preparation	Chemistry	Max Read length (bases)	Run Times (days)	Max Gb per Run
Roche 454	Clonal-emPCR	Pyrosequencing	400‡	0.42	0.40-0.60
GS FLX Titanium	Clonal-emPCR	Pyrosequencing	400‡	0.42	0.035
Illumina MiSeq	Clonal Bridge Amplification	Reversible Dye Terminator	2x300	0.17-2.7	15
Illumina HiSeq	Clonal Bridge Amplification	Reversible Dye Terminator	2x150	0.3-11	1000
Illumina Genome Analyzer IIX	Clonal Bridge Amplification	Reversible Dye Terminator	2x150	2-14	95
Life Technologies SOLiD4	Clonal-emPCR	Oligonucleotide 8-mer Chained Ligation	20-45	4-7	35-50

Life Technologies Ion Proton	Clonal-emP-CR	Native dNTPs, proton detection	200	0.5	100
Complete Genomics	Gridded DNA-nano-balls	Oligonucleotide 9-mer Unchained Ligation	7x10	11	3000
Helicos Biosciences Heliscope	Single Molecule	Reversible Dye Terminator	35‡	8	25
Pacific Biosciences SMRT	Single Molecule	Phospholinked Fluorescent Nucleotides	10,000 (N50); 30,000+ (max)	0.08	0.5

Run times and gigabase (Gb) output per run for single-end sequencing are noted. Run times and outputs approximately double when performing paired-end sequencing. ‡Average read lengths for the Roche 454 and Helicos Biosciences platforms.

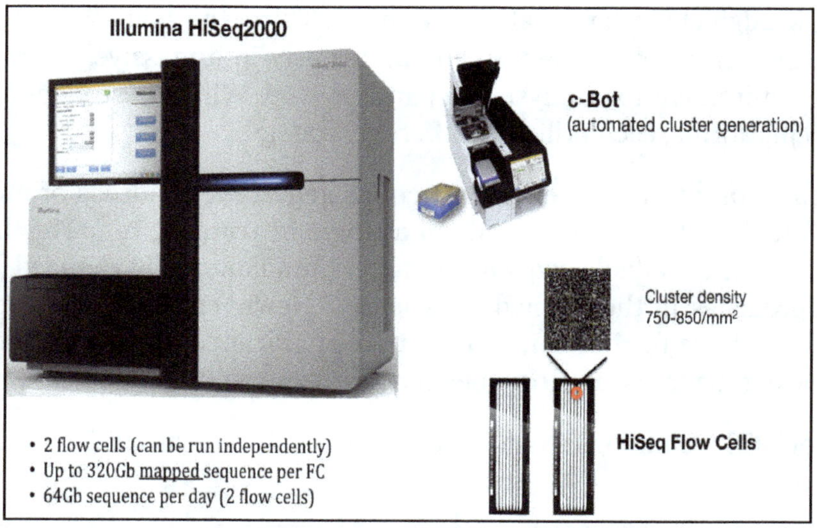

An Illumina HiSeq 2000 sequencing machine.

Template Preparation Methods for NGS

Two methods are used in preparing templates for NGS reactions: amplified templates originating from single DNA molecules, and single DNA molecule templates. For imaging systems which cannot detect single fluorescence events, amplification of DNA templates is required. The three most common amplification methods are emulsion PCR (emPCR), rolling circle and solid-phase amplification. The final distribution of templates can be spatially random or on a grid.

Emulsion PCR

In emulsion PCR methods, a DNA library is first generated through random fragmentation of genomic DNA. Single-stranded DNA fragments (templates) are attached to the surface of beads with adaptors or linkers, and one bead is attached to a single DNA fragment from the DNA library. The surface of the beads contains oligonucleotide probes with sequences that are complementary to the adaptors binding the DNA fragments. The beads are then compartmentalized into water-oil emulsion droplets. In the aqueous water-oil emulsion, each of the droplets capturing one bead is a PCR microreactor that produces amplified copies of the single DNA template.

Gridded Rolling Circle Nanoballs

Amplification of a population of single DNA molecules by rolling circle amplification in solution is followed by capture on a grid of spots sized to be smaller than the DNAs to be immobilized.

DNA Colony Generation (Bridge Amplification)

Forward and reverse primers are covalently attached at high-density to the slide in a flow cell. The ratio of the primers to the template on the support defines the surface density of the amplified clusters. The flow cell is exposed to reagents for polymerase-based extension, and priming occurs as the free/distal end of a ligated fragment "bridges" to a complementary oligo on the surface. Repeated denaturation and extension results in localized amplification of DNA fragments in millions of separate locations across the flow cell surface. Solid-phase amplification produces 100–200 million spatially separated template clusters, providing free ends to which a universal sequencing primer is then hybridized to initiate the sequencing reaction.

Single-molecule Templates

Protocols requiring DNA amplification are often cumbersome to implement and may introduce sequencing errors. The preparation of single-molecule templates is more straightforward and does not require PCR, which can introduce errors in the amplified templates. AT-rich and GC-rich target sequences often show amplification bias, which results in their underrepresentation in genome alignments and assemblies. Single molecule templates are usually immobilized on solid supports using one of at least three different approaches. In the first approach, spatially distributed individual primer molecules are covalently attached to the solid support. The template, which is prepared by randomly fragmenting the starting material into small sizes (for example,~200–250 bp) and adding common adapters to the fragment ends, is then hybridized to the immobilized primer. In the second approach, spatially distributed single-molecule templates are covalently attached to the solid support by priming and extending single-stranded, single-molecule templates from immobilized primers.

A common primer is then hybridized to the template. In either approach, DNA polymerase can bind to the immobilized primed template configuration to initiate the NGS reaction. Both of the above approaches are used by Helicos BioSciences. In a third ap proach, spatially distributed single polymerase molecules are attached to the solid support, to which a primed template molecule is bound. This approach is used by Pacific Biosciences. Larger DNA molecules (up to tens of thousands of base pairs) can be used with this technique and, unlike the first two approaches; the third approach can be used with real-time methods, resulting in potentially longer read lengths.

Uses of NGS in Clinical Practice

Clinical Genetics

There are numerous opportunities to use NGS in clinical practice to improve patient care, including:

NGS Captures a Broader Spectrum of Mutations than Sanger Sequencing

The spectrum of DNA variation in a human genome comprises small base changes (substitutions), insertions and deletions of DNA, large genomic deletions of exons or whole genes and rearrangements such as inversions and translocations. Traditional Sanger sequencing is restricted to the discovery of substitutions and small insertions and deletions. For the remaining mutations dedicated assays are frequently performed, such as fluorescence in situ hybridisation (FISH) for conventional karyotyping, or comparative genomic hybridisation (CGH) microarrays to detect submicroscopic chromosomal copy number changes such as microdeletions. However, these data can also be derived from NGS sequencing data directly, obviating the need for dedicated assays while harvesting the full spectrum of genomic variation in a single experiment. The only limitations reside in regions which sequence poorly or map erroneously due to extreme guanine/cytosine (GC) content or repeat architecture, for example, the repeat expansions underlying Fragile X syndrome, or Huntington's disease.

Genomes can be Interrogated without Bias

Capillary sequencing depends on preknowledge of the gene or locus under investigation. However, NGS is completely unselective and used to interrogate full genomes or exomes to discover entirely novel mutations and disease causing genes. In paediatrics, this could be exploited to unravel the genetic basis of unexplained syndromes. For example, a nationwide project, Deciphering Developmental Disorders, running at the Wellcome Trust Sanger Institute in collaboration with NHS clinical genetics services aims to unravel the genetic basis of unexplained developmental delay by sequencing affected children and their parents to uncover deleterious de novo variants. Allying these molecular data with detailed clinical phenotypic information has been successful in identifying novel genes mutated in affected children with similar clinical features.

Increased Sensitivity of NGS Allows Detection of Mosaic Mutations

Mosaic mutations are acquired as a postfertilisation event and consequently they present at variable frequency within the cells and tissues of an individual. Capillary sequencing may miss these variants as they frequently present with a subtlety which falls below the sensitivity of the technology. NGS sequencing provides a far more sensitive read-out and can therefore be used to identify variants which reside in just a few per cent of the cells, including mosaic variation. In addition, the sensitivity of NGS sequencing can be increased further, simply by increasing sequencing depth. This has seen NGS employed for very sensitive investigations such as interrogating foetal DNA from maternal blood or tracking the levels of tumour cells from the circulation of cancer patients.

Limitations

The main disadvantage of NGS in the clinical setting is putting in place the required infrastructure, such as computer capacity and storage, and also the personnel expertise required to comprehensively analyse and interpret the subsequent data. In addition, the volume of data needs to be managed skilfully to extract the clinically important information in a clear and robust interface. The actual sequencing cost of NGS is negligible.

For example, a state of the art NGS platform can generate approximately 150 000 000 reads for around £1000 whereas a single Sanger read typically costs less than £1. However, to make NGS cost effective one would have to run large batches of samples which may require supra-regional centralisation. Following the initial capital investment, the capacity of a NGS facility can provide a service on national scale likely offering economic benefits in addition to improvements in patient care.

SANGER SEQUENCING

Sanger sequencing is a method of DNA sequencing based on the selective incorporation of chain-terminating dideoxynucleotides by DNA polymerase during in vitro DNA replication. Developed by Frederick Sanger and colleagues in 1977, it was the most widely used sequencing method for approximately 40 years. It was first commercialized by Applied Biosystems in 1986. More recently, higher volume Sanger sequencing has been replaced by "Next-Gen" sequencing methods, especially for large-scale, automated genome analyses.

However, the Sanger method remains in wide use, for smaller-scale projects, and for validation of Next-Gen results. It still has the advantage over short-read sequencing technologies (like Illumina) that it can produce DNA sequence reads of > 500 nucleotides.

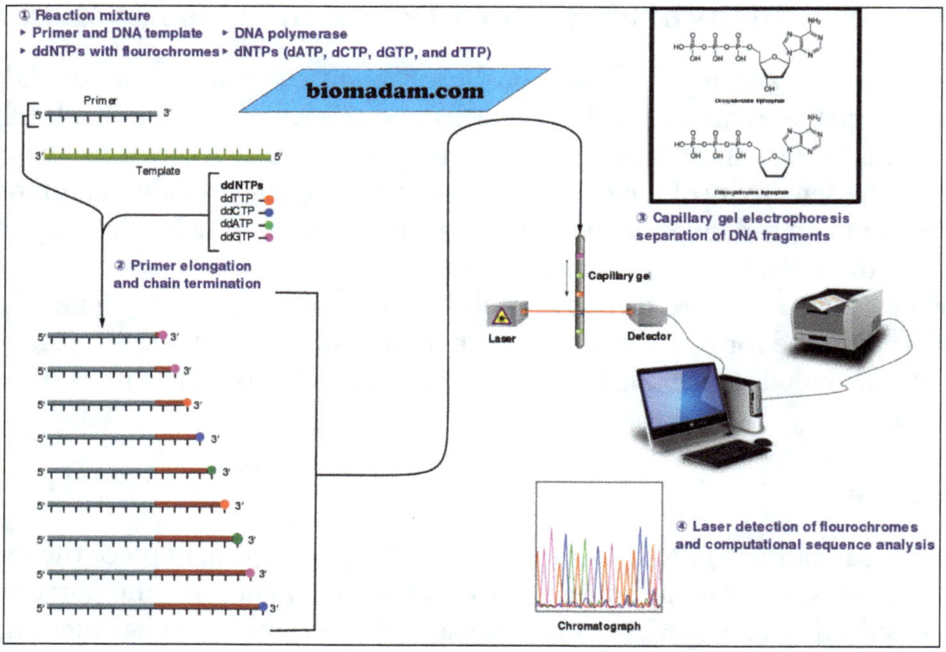

The Sanger (chain-termination) method for DNA sequencing.

Method

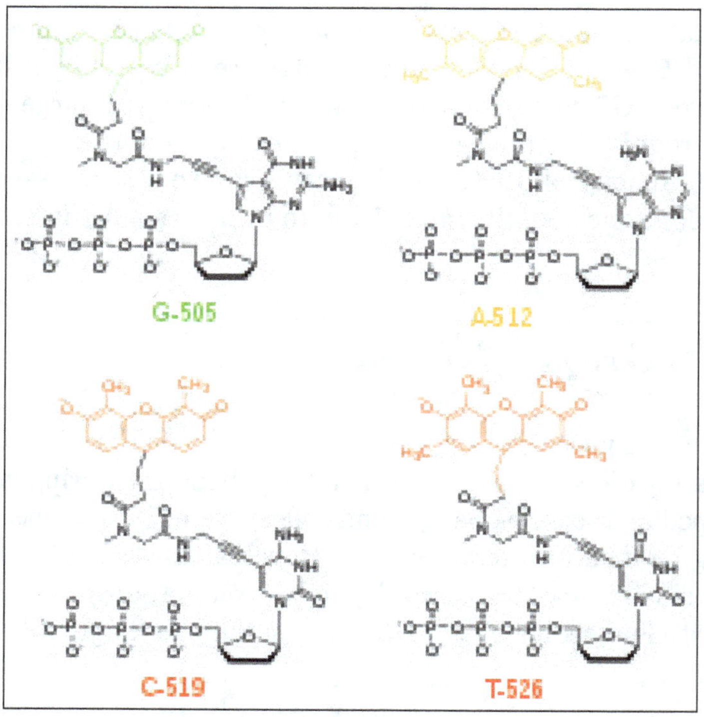

Fluorescent ddNTP molecules.

The classical chain-termination method requires a single-stranded DNA template, a DNA primer, a DNA polymerase, normal deoxynucleotidetriphosphates (dNTPs), and modified di-deoxynucleotidetriphosphates (ddNTPs), the latter of which terminate DNA strand elongation. These chain-terminating nucleotides lack a 3'-OH group required for the formation of a phosphodiester bond between two nucleotides, causing DNA polymerase to cease extension of DNA when a modified ddNTP is incorporated. The ddNTPs may be radioactively or fluorescently labelled for detection in automated sequencing machines.

The DNA sample is divided into four separate sequencing reactions, containing all four of the standard deoxynucleotides (dATP, dGTP, dCTP and dTTP) and the DNA polymerase. To each reaction is added only one of the four dideoxynucleotides (ddATP, ddGTP, ddCTP, or ddTTP), while the other added nucleotides are ordinary ones. The dideoxynucleotide concentration should be approximately 100-fold higher than that of the corresponding deoxynucleotide (e.g. 0.5mM ddTTP : 0.005mM dTTP) to allow enough fragments to be produced while still transcribing the complete sequence. Putting it in a more sensible order, four separate reactions are needed in this process to test all four ddNTPs. Following rounds of template DNA extension from the bound primer, the resulting DNA fragments are heat denatured and separated by size using gel electrophoresis. In the original publication of 1977, the formation of base-paired loops of ssDNA was a cause of serious difficulty in resolving bands at some locations. This is frequently performed using a denaturing polyacrylamide-urea gel with each of the four reactions run in one of four individual lanes (lanes A, T, G, C). The DNA bands may then be visualized by autoradiography or UV light and the DNA sequence can be directly read off the X-ray film or gel image.

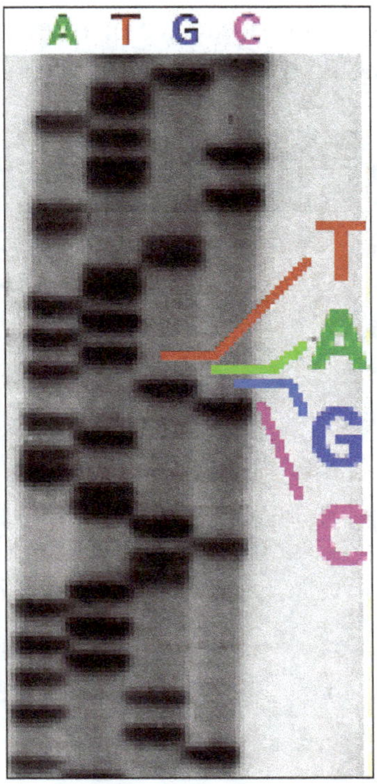

Part of a radioactively labelled sequencing gel.

In the image on the right, X-ray film was exposed to the gel, and the dark bands correspond to DNA fragments of different lengths. A dark band in a lane indicates a DNA fragment that is the result of chain termination after incorporation of a dideoxynucleotide (ddATP, ddGTP, ddCTP, or ddTTP). The relative positions of the different bands among the four lanes, from bottom to top, are then used to read the DNA sequence.

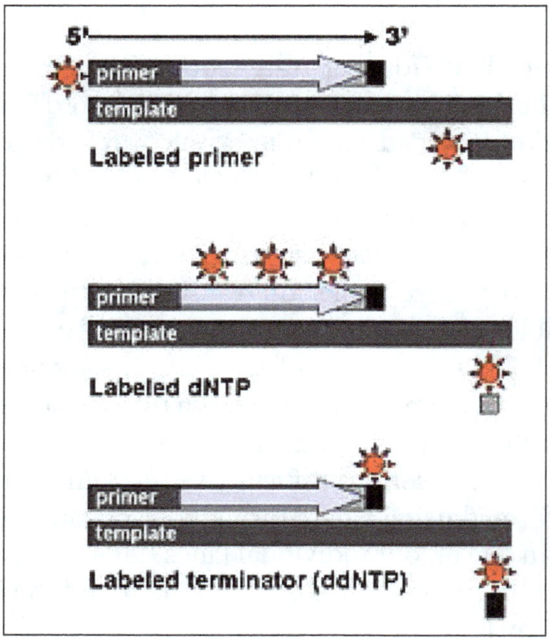

DNA fragments are labelled with a radioactive or fluorescent tag on the primer (1), in the new DNA strand with a labeled dNTP, or with a labeled ddNTP.

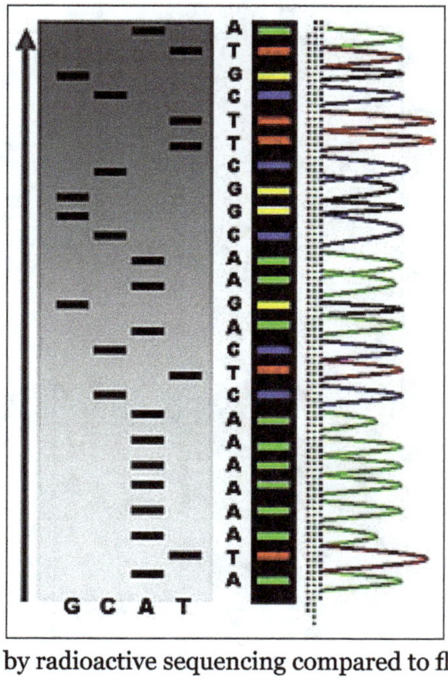

Sequence ladder by radioactive sequencing compared to fluorescent peaks.

Technical variations of chain-termination sequencing include tagging with nucleotides containing radioactive phosphorus for radiolabelling, or using a primer labeled at the 5' end with a fluorescent dye. Dye-primer sequencing facilitates reading in an optical system for faster and more economical analysis and automation. The later development by Leroy Hood and coworkers of fluorescently labeled ddNTPs and primers set the stage for automated, high-throughput DNA sequencing.

Chain-termination methods have greatly simplified DNA sequencing. For example, chain-termination-based kits are commercially available that contain the reagents needed for sequencing, pre-aliquoted and ready to use. Limitations include non-specific binding of the primer to the DNA, affecting accurate read-out of the DNA sequence, and DNA secondary structures affecting the fidelity of the sequence.

Dye-terminator Sequencing

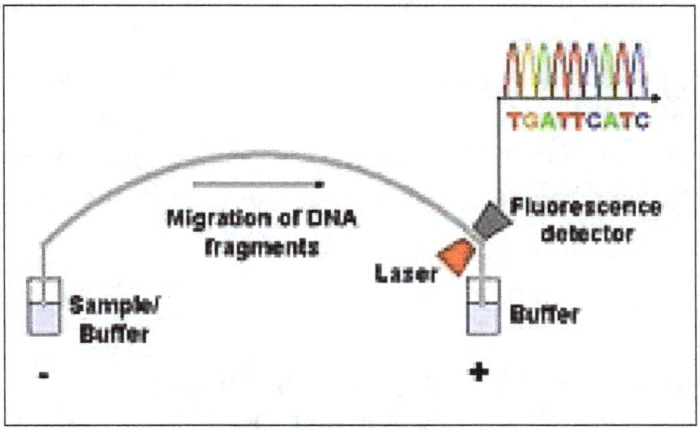

Capillary electrophoresis.

Dye-terminator sequencing utilizes labelling of the chain terminator ddNTPs, which permits sequencing in a single reaction, rather than four reactions as in the labelled-primer method. In dye-terminator sequencing, each of the four dideoxynucleotide chain terminators is labelled with fluorescent dyes, each of which emit light at different wavelengths.

Owing to its greater expediency and speed, dye-terminator sequencing is now the mainstay in automated sequencing. Its limitations include dye effects due to differences in the incorporation of the dye-labelled chain terminators into the DNA fragment, resulting in unequal peak heights and shapes in the electronic DNA sequence trace chromatogram after capillary electrophoresis.

This problem has been addressed with the use of modified DNA polymerase enzyme systems and dyes that minimize incorporation variability, as well as methods for eliminating "dye blobs". The dye-terminator sequencing method, along with automated high-throughput DNA sequence analyzers, was used for the vast majority of sequencing projects until the introduction of Next Generation Sequencing.

Automation and Sample Preparation

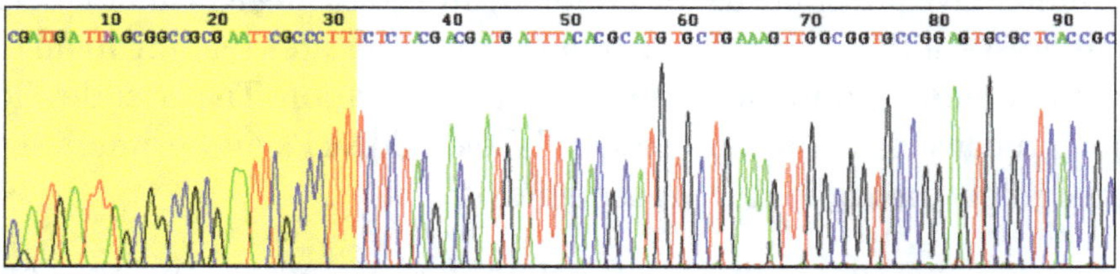

View of the start of an example dye-terminator read.

Automated DNA-sequencing instruments (DNA sequencers) can sequence up to 384 DNA samples in a single batch. Batch runs may occur up to 24 times a day. DNA sequencers separate strands by size (or length) using capillary electrophoresis, they detect and record dye fluorescence, and output data as fluorescent peak trace chromatograms. Sequencing reactions (thermocycling and labelling), cleanup and re-suspension of samples in a buffer solution are performed separately, before loading samples onto the sequencer. A number of commercial and non-commercial software packages can trim low-quality DNA traces automatically. These programs score the quality of each peak and remove low-quality base peaks (which are generally located at the ends of the sequence). The accuracy of such algorithms is inferior to visual examination by a human operator, but is adequate for automated processing of large sequence data sets.

Challenges

Common challenges of DNA sequencing with the Sanger method include poor quality in the first 15-40 bases of the sequence due to primer binding and deteriorating quality of sequencing traces after 700-900 bases. Base calling software such as Phred typically provides an estimate of quality to aid in trimming of low-quality regions of sequences.

In cases where DNA fragments are cloned before sequencing, the resulting sequence may contain parts of the cloning vector. In contrast, PCR-based cloning and next-generation sequencing technologies based on pyrosequencing often avoid using cloning vectors. Recently, one-step Sanger sequencing (combined amplification and sequencing) methods such as Ampliseq and SeqSharp have been developed that allow rapid sequencing of target genes without cloning or prior amplification.

Current methods can directly sequence only relatively short (300-1000 nucleotides long) DNA fragments in a single reaction. The main obstacle to sequencing DNA fragments above this size limit is insufficient power of separation for resolving large DNA fragments that differ in length by only one nucleotide.

Microfluidic Sanger Sequencing

Microfluidic Sanger sequencing is a lab-on-a-chip application for DNA sequencing, in which the Sanger sequencing steps (thermal cycling, sample purification, and capillary electrophoresis) are integrated on a wafer-scale chip using nanoliter-scale sample volumes. This technology generates long and accurate sequence reads, while obviating many of the significant shortcomings of the conventional Sanger method (e.g. high consumption of expensive reagents, reliance on expensive equipment, personnel-intensive manipulations, etc.) by integrating and automating the Sanger sequencing steps.

In its modern inception, high-throughput genome sequencing involves fragmenting the genome into small single-stranded pieces, followed by amplification of the fragments by Polymerase Chain Reaction (PCR). Adopting the Sanger method, each DNA fragment is irreversibly terminated with the incorporation of a fluorescently labeled dideoxy chain-terminating nucleotide, thereby producing a DNA "ladder" of fragments that each differ in length by one base and bear a base-specific fluorescent label at the terminal base. Amplified base ladders are then separated by Capillary Array Electrophoresis (CAE) with automated, *in situ* "finish-line" detection of the fluorescently labeled ssDNA fragments, which provides an ordered sequence of the fragments. These sequence reads are then computer assembled into overlapping or contiguous sequences (termed "contigs") which resemble the full genomic sequence once fully assembled.

Sanger methods achieve read lengths of approximately 800bp (typically 500-600bp with non-enriched DNA). The longer read lengths in Sanger methods display significant advantages over other sequencing methods especially in terms of sequencing repetitive regions of the genome. A challenge of short-read sequence data is particularly an issue in sequencing new genomes *(de novo)* and in sequencing highly rearranged genome segments, typically those seen of cancer genomes or in regions of chromosomes that exhibit structural variation.

Applications of Microfluidic Sequencing Technologies

Other useful applications of DNA sequencing include single nucleotide polymorphism (SNP) detection, single-strand conformation polymorphism (SSCP) heteroduplex analysis, and short tandem repeat (STR) analysis. Resolving DNA fragments according to differences in size and conformation is the most critical step in studying these features of the genome.

Device Design

The sequencing chip has a four-layer construction, consisting of three 100-mm-diameter glass wafers (on which device elements are microfabricated) and a polydimethylsiloxane (PDMS) membrane. Reaction chambers and capillary electrophoresis channels are etched between the top two glass wafers, which are thermally bonded. Three-dimensional channel interconnections and microvalves are formed by the PDMS and bottom manifold glass wafer.

The device consists of three functional units, each corresponding to the Sanger sequencing steps. The Thermal Cycling (TC) unit is a 250-nanoliter reaction chamber with integrated resistive temperature detector, microvalves, and a surface heater. Movement of reagent between the top all-glass layer and the lower glass-PDMS layer occurs through 500-μm-diameter via-holes. After thermal-cycling, the reaction mixture undergoes purification in the capture/purification chamber, and then is injected into the capillary electrophoresis (CE) chamber. The CE unit consists of a 30-cm capillary which is folded into a compact switchback pattern via 65-μm-wide turns.

Sequencing Chemistry

- Thermal cycling: In the TC reaction chamber, dye-terminator sequencing reagent, template DNA, and primers are loaded into the TC chamber and thermal-cycled for 35 cycles (at 95 °C for 12 seconds and at 60 °C for 55 seconds).
- Purification: The charged reaction mixture (containing extension fragments, template DNA, and excess sequencing reagent) is conducted through a capture/purification chamber at 30 °C via a 33-Volts/cm electric field applied between capture outlet and inlet ports. The capture gel through which the sample is driven, consists of 40 μM of oligonucleotide (complementary to the primers) covalently bound to a polyacrylamide matrix. Extension fragments are immobilized by the gel matrix, and excess primer, template, free nucleotides, and salts are eluted through the capture waste port. The capture gel is heated to 67-75 °C to release extension fragments.
- Capillary electrophoresis: Extension fragments are injected into the CE chamber where they are electrophoresed through a 125-167-V/cm field.

Platforms

The Apollo 100 platform integrates the first two Sanger sequencing steps (thermal cycling and purification) in a fully automated system. The manufacturer claims that samples are ready for capillary electrophoresis within three hours of the sample and reagents being loaded into the system. The Apollo 100 platform requires sub-microliter volumes of reagents.

PYROSEQUENCING

Pyrosequencing is a method of DNA sequencing (determining the order of nucleotides in DNA) based on the "sequencing by synthesis" principle, in which the sequencing is performed by detecting the nucleotide incorporated by a DNA polymerase.

Pyrosequencing relies on light detection based on a chain reaction when pyrophosphate is released. Hence, the name pyrosequencing.

The principle of Pyrosequencing was first described in 1993 by Bertil Pettersson, Mathias Uhlen and Pål Nyren by combining the solid phase sequencing method using streptavidin coated magnetic beads with recombinant DNA polymerase lacking 3´to 5´exonuclease activity (proof-reading) and luminescence detection using the firefly luciferase enzyme. A mixture of three enzymes (DNA polymerase, ATP sulfurylase and firefly luciferase) and a nucleotide (dNTP) are added to single stranded DNA to be sequenced and the incorporation of nucleotide is followed by measuring the light emitted. The intensity of the light determines if 0, 1 or more nucleotides have been incorporated, thus showing how many complementary nucleotides are present on the template strand. The nucleotide mixture is removed before the next nucleotide mixture is added. This process is repeated with each of the four nucleotides until the DNA sequence of the single stranded template is determined.

A second solution-based method for Pyrosequencing was described in 1998 by Mostafa Ronaghi, Mathias Uhlen and Pål Nyren. In this alternative method, an additional enzyme apyrase is introduced to remove nucleotides that are not incorporated by the DNA polymerase. This enabled the enzyme mixture including the DNA polymerase, the luciferase and the apyrase to be added at the start and kept throughout the procedure, thus providing a simple set-up suitable for automation. An automated instrument based on this principle was introduced to the market the following year by the company Pyrosequencing.

A third microfluidic variant of the Pyrosequencing method was described in 2005 by Jonathan Rothberg and co-workers at the company 454 Life Sciences. This alternative approach for Pyrosequencing was based on the original principle of attaching the DNA to be sequenced to a solid support and they showed that sequencing could be performed in a highly parallel manner using a microfabricated microarray. This allowed for high-throughput DNA sequencing and an automated instrument was introduced to the market. This became the first next generation sequencing instrument starting a new era in genomics research, with rapidly falling prices for DNA sequencing allowing whole genome sequencing at affordable prices.

Procedure

"Sequencing by synthesis" involves taking a single strand of the DNA to be sequenced and then synthesizing its complementary strand enzymatically. The pyrosequencing method is based on detecting the activity of DNA polymerase (a DNA synthesizing enzyme) with another chemoluminescent enzyme. Essentially, the method allows sequencing a single strand of DNA by synthesizing the complementary strand along it, one base pair at a time, and detecting which base was actually added at each step. The template DNA is immobile, and solutions of A, C, G, and T nucleotides are sequentially added and removed from the reaction. Light is produced only when the nucleotide solution complements the first unpaired base of the template. The sequence of solutions which produce chemiluminescent signals allows the determination of the sequence of the template.

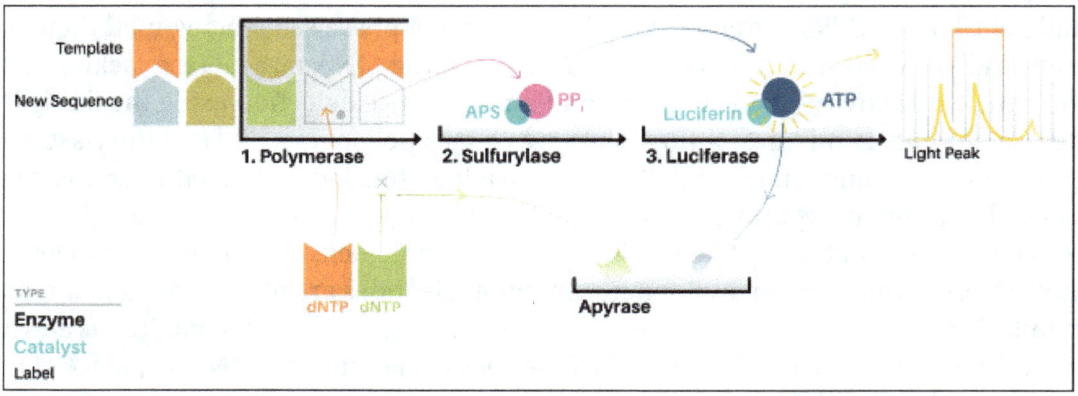

The chart shows how pyrosequencing works.

For the solution-based version of Pyrosequencing, the single-strand DNA (ssDNA) template is hybridized to a sequencing primer and incubated with the enzymes DNA polymerase, ATP sulfurylase, luciferase and apyrase, and with the substrates adenosine 5´ phosphosulfate (APS) and luciferin.

- The addition of one of the four deoxynucleotide triphosphates (dNTPs) (dATPαS, which is not a substrate for a luciferase, is added instead of dATP to avoid noise) initiates the second step. DNA polymerase incorporates the correct, complementary dNTPs onto the template. This incorporation releases pyrophosphate (PPi).

- ATP sulfurylase converts PPi to ATP in the presence of adenosine 5´ phosphosulfate. This ATP acts as a substrate for the luciferase-mediated conversion of luciferin to oxyluciferin that generates visible light in amounts that are proportional to the amount. The light produced in the luciferase-catalyzed reaction is detected by a camera and analyzed in a program.

- Unincorporated nucleotides and ATP are degraded by the apyrase, and the reaction can restart with another nucleotide.

Limitations

Currently, a limitation of the method is that the lengths of individual reads of DNA sequence are in the neighborhood of 300-500 nucleotides, shorter than the 800-1000 obtainable with chain termination methods (e.g. Sanger sequencing).

This can make the process of genome assembly more difficult, particularly for sequences containing a large amount of repetitive DNA. Lack of proof-reading activity limits accuracy of this method.

ILLUMINA DYE SEQUENCING

Illumina dye sequencing is a technique used to determine the series of base pairs in DNA, also known as DNA sequencing. The reversible terminated chemistry concept was invented by Bruno Canard and Simon Sarfati at the Pasteur Institute in Paris. It was developed by Shankar Balasubramanian and David Klenerman of Cambridge University, who subsequently founded Solexa, a company later acquired by Illumina. This sequencing method is based on reversible dye-terminators that enable the identification of single bases as they are introduced into DNA strands. It can also be used for whole-genome and region sequencing, transcriptome analysis, metagenomics, small RNA discovery, methylation profiling, and genome-wide protein-nucleic acid interaction analysis.

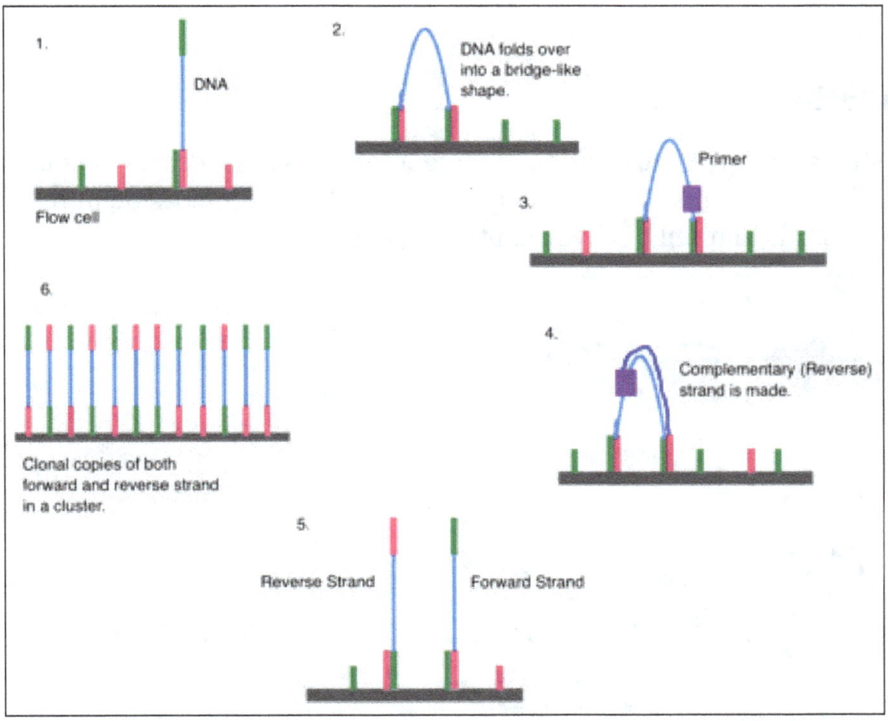

The DNA attaches to the flow cell via complementary sequences. The strand bends over and attaches to a second oligo forming a bridge. A polymerase synthesizes the reverse strand. The two strands release and straighten. Each forms a new bridge (bridge amplification). The result is a cluster of DNA forward and reverse strands clones.

Illumina sequencing technology works in three basic steps: amplify, sequence, and analyze. The process begins with purified DNA. The DNA gets chopped up into smaller pieces and given adapters, indices, and other kinds of molecular modifications that act as reference points during amplification, sequencing, and analysis. The modified DNA is loaded onto a specialized chip where amplification and sequencing will take place.

Along the bottom of the chip are hundreds of thousands of oligonucleotides (short, synthetic pieces of DNA). They are anchored to the chip and able to grab DNA fragments that have complementary sequences. Once the fragments have attached, a phase called cluster generation begins. This step makes about a thousand copies of each fragment of DNA. Next, primers and modified nucleotides enter the chip. These nucleotides have reversible 3' blockers that force the polymerase to add on only one nucleotide at a time as well as fluorescent tags. After each round of synthesis, a camera takes a picture of the chip. A computer determines what base was added by the wavelength of the fluorescent tag and records it for every spot on the chip. After each round, non-incorporated molecules are washed away. A chemical deblocking step is then used in the removal of the 3' terminal blocking group and the dye in a single step. The process continues until the full DNA molecule is sequenced. With this technology, thousands of places throughout the genome are sequenced at once via massive parallel sequencing.

Procedure

Tagmentation

The first step after DNA purification is tagmentation. Transposases randomly cut the DNA into short segments ("tags"). Adapters are added on either side of the cut points (ligation). Strands that fail to have adapters ligated are washed away.

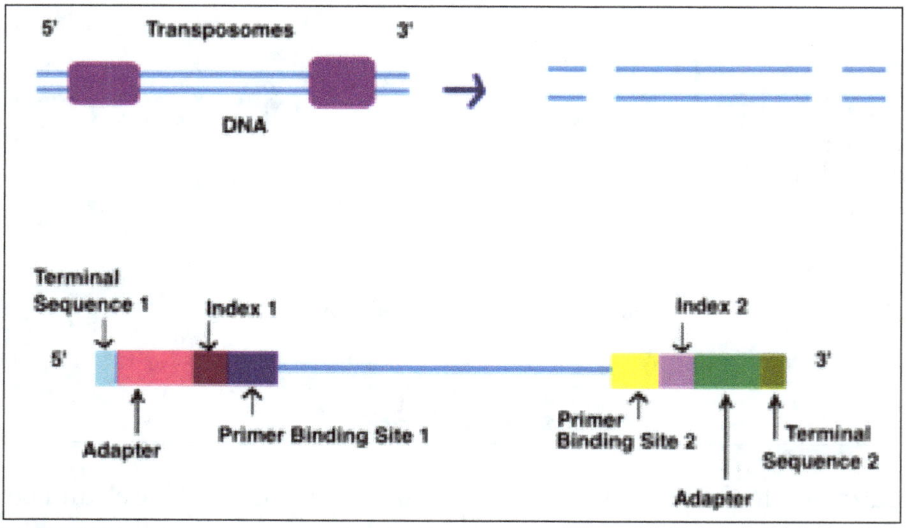

Double stranded DNA is cleaved by transposomes. The cut ends are repaired and adapters, indices, primer binding sites, and terminal sites are added to each strand of the DNA.

Reduced Cycle Amplification

The next step is called reduced cycle amplification. During this step, sequences for primer binding, indices, and terminal sequences are added. Indices are usually six base pairs long and are used during DNA sequence analysis to identify samples. Indices allow for up to 96 different samples to be run together. During analysis, the computer will group all reads with the same index together. The terminal sequences are used for attaching the DNA strand to the flow cell. Illumina uses a "sequence by synthesis" approach. This process takes place inside of an acrylamide-coated glass flow cell. The flow cell has oligonucleotides (short nucleotide sequences) coating the bottom of the cell, and they serve to hold the DNA strands in place during sequencing. The oligos match the two kinds of terminal sequences added to the DNA during reduced cycle amplification. As the DNA enters the flow cell, one of the adapters attaches to a complementary oligo.

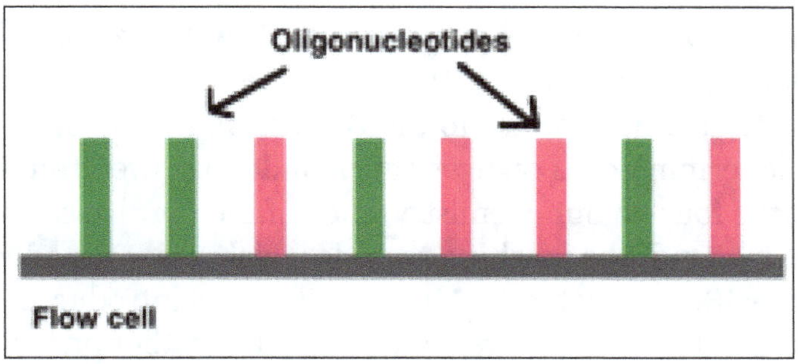

Millions of oligos line the bottom of each flow cell lane.

Bridge Amplification

Once attached, cluster generation can begin. The goal is to create hundreds of identical strands of DNA. Some will be the forward strand; the rest, the reverse. Clusters are generated through bridge amplification. Polymerases move along a strand of DNA, creating its complementary strand. The original strand is washed away, leaving only the reverse strand. At the top of the reverse strand there is an adapter sequence. The DNA strand bends and attaches to the oligo that is complementary to the top adapter sequence. Polymerases attach to the reverse strand, and its complementary strand (which is identical to the original) is made. The now double stranded DNA is denatured so that each strand can separately attach to an oligonucleotide sequence anchored to the flow cell. One will be the reverse strand; the other, the forward. This process is called bridge amplification, and it happens for thousands of clusters all over the flow cell at once.

Clonal Amplification

Over and over again, DNA strands will bend and attach to oligos. Polymerases will synthesize a new strand to create a double stranded segment, and that will be denatured so that all of the DNA strands in one area are from a single source (clonal amplification). Clonal amplification is important for quality control purposes. If a strand is found to have an odd sequence, then scientists can check the reverse strand to make sure that it has the complement of the same oddity. The forward and reverse strands act as checks to guard against artifacts. Because Illumina sequencing uses polymerases, base substitution errors have been observed, especially at the 3' end. Paired end reads combined with cluster generation can confirm an error took place. The reverse and forward strands should be complementary to each other, all reverse reads should match each other, and all forward reads should match each other. If a read is not similar enough to its counterparts (with which it should be a clone), an error may have occurred. A minimum threshold of 97% similarity has been used in some labs' analyses.

Sequence by Synthesis

At the end of clonal amplification, all of the reverse strands are washed off the flow cell, leaving only forward strands. Primers attach to the forward strands and a polymerase adds fluorescently tagged nucleotides to the DNA strand. Only one base is added per round. A reversible terminator is on every nucleotide to prevent multiple additions in one round. Using the four-colour chemistry, each of the four bases has a unique emission, and after each round, the machine records which base was added. Starting with the launch of the NextSeq and later the MiniSeq, Illumina introduced a new two-colour sequencing chemistry. Nucleotides are distinguished by either one of two colours (red or green), no colour ("black") or binding both colours (appearing orange as a mixture between red and green).

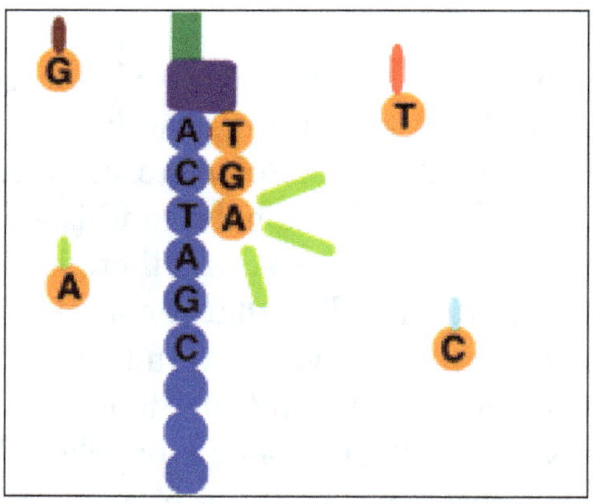

Tagged nucleotides are added in order to the DNA strand. Each of the four nucleotides have an identifying label that can be excited to emit a characteristic wavelength. A computer records all of the emissions, and from this data, base calls are made.

Once the DNA strand has been read, the strand that was just added is washed away. Then, the index 1 primer attaches, polymerizes the index 1 sequence, and is washed away. The strand forms a bridge again, and the 3' end of the DNA strand attaches to an oligo on the flow cell. The index 2 primer attaches, polymerizes the sequence, and is washed away.

A polymerase sequences the complementary strand on top of the arched strand. They separate, and the 3' end of each strand is blocked. The forward strand is washed away, and the process of sequence by synthesis repeats for the reverse strand.

Data Analysis

The sequencing occurs for millions of clusters at once, and each cluster has ~1,000 identical copies of a DNA insert. The sequence data is analyzed by finding fragments with overlapping areas, called contigs, and lining them up. If a reference sequence is known, the contigs are then compared to it for variant identification.

This piecemeal process allows scientists to see the complete sequence even though an unfragmented sequence was never run; however, because Illumina read lengths are not very long (HiSeq sequencing can produce read lengths around 90 bp long), it can be a struggle to resolve short tandem repeat areas. Also, if the sequence is de novo and so a reference doesn't exist, repeated areas can cause a lot of difficulty in sequence assembly. Additional difficulties include base substitutions (especially at the 3' end of reads) by inaccurate polymerases, chimeric sequences, and PCR-bias, all of which can contribute to generating an incorrect sequence.

Comparison with other Sequencing Methods

This technique offers a number of advantages over traditional sequencing methods such as Sanger sequencing. Due to the automated nature of Illumina dye sequencing it is possible to sequence multiple strands at once and gain actual sequencing data quickly. Additionally, this method only uses DNA polymerase as opposed to multiple, expensive enzymes required by other sequencing techniques (i.e. pyrosequencing).

Examples of Use

Illumina sequencing has been used to research transcriptomes of the sweet potato and the gymnosperm genus *Taxus*.

SOLID

SOLiD (Sequencing by Oligonucleotide Ligation and Detection) is a next-generation DNA sequencing technology developed by Life Technologies and has been commercially available since 2006. This next generation technology generates hundreds of millions to billions of small sequence reads at one time.

This method should not be confused with "sequencing by synthesis," a principle used by Roche-454 pyrosequencing, and the Solexa system.

The method used by this machine (sequencing-by-ligation) has been reported to have some issue sequencing palindromic sequences.

Chemistry

A library of DNA fragments is prepared from the sample to be sequenced, and is used to prepare clonal bead populations. That is, only one species of fragment will be present on the surface of each magnetic bead. The fragments attached to the magnetic beads will have a universal P1 adapter sequence attached so that the starting sequence of every fragment is both known and identical. Emulsion PCR takes place in microreactors containing all the necessary reagents for PCR. The resulting PCR products attached to the beads are then covalently bound to a glass slide.

Primers hybridize to the P1 adapter sequence within the library template. A set of four fluorescently labelled di-base probes compete for ligation to the sequencing primer. Specificity of the di-base probe is achieved by interrogating every 1st and 2nd base in each ligation reaction. Multiple cycles of ligation, detection and cleavage are performed with the number of cycles determining the eventual read length. Following a series of ligation cycles, the extension product is removed and the template is reset with a primer complementary to the n-1 position for a second round of ligation cycles.

Five rounds of primer reset are completed for each sequence tag. Through the primer reset process, each base is interrogated in two independent ligation reactions by two different primers. For example, the base at read position 5 is assayed by primer number 2 in ligation cycle 2 and by primer number 3 in ligation cycle 1.

Throughput and Accuracy

The SOLiD 3plus platform yields 60 gigabases of usable DNA data per run. Due to the two base encoding system, an inherent accuracy check is built into the technology and offers 99.94% accuracy. The chemistry of the systems also means that it is not hindered by homopolymers unlike the Roche 454 FLX system and so large and difficult homopolymer repeat regions are no longer a problem to sequence.

Applications

Naturally the technology will be used to sequence DNA, but because of the high parallel nature of all next generation technologies they also have applications in transcriptomics and epigenomics.

Microarrays was once the mainstay of the transcriptomics the last ten years and array based technology has subsequently branched out to other areas. However, they are limited in that only information can be obtained for probes that are on the chip. Only information for organisms for which chips are available can obtained, and they come with all the problems of hybridizing large numbers of molecules (differing hybridizing temperatures). RNA-Seq transcriptomics by next gen sequencing will mean these barriers no longer hold true. Any organism's entire transcriptome could be potentially sequenced in one run (for very small bacterial genomes) and not only would the identification of each transcript be available but expression profiling is possible as quantitative reads can also be achieved.

Chromatin immunoprecipitation (ChIP) is a method for determining transcription factor binding sites and DNA-protein interactions. It has in the past been combined with array technology (ChIP-chip) with some success. Next gen sequencing can also be applied in this area. Methylation immunoprecipitation (MeDIP) can also be performed and also on arrays.

The ability to learn more about methylation and TF binding sites on a genome wide scale is a valuable resource and could teach us much about disease and molecular biology in general.

THIRD-GENERATION SEQUENCING

Third-generation sequencing (also known as long-read sequencing) is a class of DNA sequencing methods currently under active development. Third generation sequencing works by reading the nucleotide sequences at the single molecule level, in contrast to existing methods that require breaking long strands of DNA into small segments then inferring nucleotide sequences by amplification and synthesis. Critical challenges exist in the engineering of the necessary molecular instruments for whole genome sequencing to make the technology commercially available.

Second-generation sequencing (short-read sequencing), often referred to as Next-generation sequencing (NGS), has dominated the DNA sequencing space since its development. It has dramatically reduced the cost of DNA sequencing by enabling a massively-paralleled approach capable of producing large numbers of reads at exceptionally high coverages throughout the genome.

Since eukaryotic genomes contain many repeated sequence (DNA), a major limitation to this class of sequencing methods is the length of reads it produces. Briefly, second generation sequencing works by first amplifying the DNA molecule and then conducting sequencing by synthesis. The collective fluorescent signal resulting from synthesizing a large number of amplified identical DNA strands allows the inference of nucleotide identity. However, due to random errors, DNA synthesis between the amplified DNA strands would become progressively out-of-sync. Quickly, the signal quality deteriorates as the read-length grows. In order to preserve read quality, long DNA molecules must be broken up into small segments, resulting in a critical limitation of second generation sequencing technologies. Computational efforts aimed to overcome this challenge often rely on approximative heuristics that may not result in accurate assemblies.

By enabling direct sequencing of single DNA molecules, third generation sequencing technologies have the capability to produce substantially longer reads than second generation sequencing. Such an advantage has critical implications for both genome science and the study of biology in general. However, third generation sequencing data have much higher error rates than previous technologies, which can complicate downstream genome assembly and analysis of the resulting data. These technologies are undergoing active development and it is expected that there will be improvements to the high error rates. For applications that are more tolerant to error rates, such as structural variant calling, third generation sequencing has been found to outperform existing methods.

Current Technologies

Sequencing technologies with a different approach than second-generation platforms were first described as "third-generation" in 2008-2009. There are several companies currently at the heart of third generation sequencing technology development, namely, Pacific Biosciences, Oxford Nanopore Technology, Quantapore (CA-USA), and Stratos (WA-USA). These companies are taking fundamentally different approaches to sequencing single DNA molecules.

PacBio developed the sequencing platform of single molecule real time sequencing (SMRT), based on the properties of zero-mode waveguides. Signals are in the form of fluorescent light emission from each nucleotide incorporated by a DNA polymerase bound to the bottom of the zL well.

Oxford Nanopore's technology involves passing a DNA molecule through a nanoscale pore structure and then measuring changes in electrical field surrounding the pore; while Quantapore has a different proprietary nanopore approach. Stratos Genomics spaces out the DNA bases with polymeric inserts, "*Xpandomers*", to circumvent the signal to noise challenge of nanopore ssDNA reading. Also notable is Helicos's single molecule fluorescence approach, but the company entered bankruptcy in the fall of 2015.

Advantages

Longer Reads

In comparison to the current generation of sequencing technologies, third generation sequencing has the obvious advantage of producing much longer reads. It is expected that these longer read lengths will alleviate numerous computational challenges surrounding genome assembly, transcript reconstruction, and metagenomics among other important areas of modern biology and medicine.

It is well known that eukaryotic genomes including primates and humans are complex and have large numbers of long repeated regions. Short reads from second generation sequencing must resort to approximative strategies in order to infer sequences over long ranges for assembly and genetic variant calling. Pair end reads have been leveraged by second generation sequencing to combat these limitations. However, exact fragment lengths of pair ends are often unknown and must also be approximated as well. By making long reads lengths possible, third generation sequencing technologies have clear advantages.

Epigenetics

Epigenetic markers are stable and potentially heritable modifications to the DNA molecule that are not in its sequence. An example is DNA methylation at CpG sites, which has been found to influence gene expression. Histone modifications are another example. The current generation of sequencing technologies rely on laboratory techniques such as ChIP-sequencing for the detection of epigenetic markers. These techniques involve tagging the DNA strand, breaking and filtering fragments that contain markers, followed by sequencing. Third generation sequencing may enable direct detection of these markers due to their distinctive signal from the other four nucleotide bases.

Portability and Speed

Other important advantages of third generation sequencing technologies include portability and sequencing speed. Since minimal sample preprocessing is required in comparison to second generation sequencing, smaller equipments could be designed. Oxford Nanopore Technology has recently commercialized the MinION sequencer. This sequencing machine is roughly the size of a regular USB flash drive and can be used readily by connecting to a laptop. In addition, since the sequencing process is not parallelized across regions of the genome, data could be collected and analyzed in real time. These advantages of third generation sequencing may be well-suited in hospital settings where quick and on-site data collection and analysis is demanded.

Challenges

Third generation sequencing, as it currently stands, faces important challenges mainly surrounding accurate identification of nucleotide bases; error rates are still much higher compared to second generation sequencing. This is generally due to instability of the molecular machinery involved. For example, in PacBio's single molecular and real time sequencing technology, the DNA polymerase molecule becomes increasingly damaged as the sequencing process occurs. Additionally, since the process happens quickly, the signals given off by individual bases may be blurred by signals from neighbouring bases. This poses a new computational challenge for deciphering the signals and consequently inferring the sequence. Methods such as Hidden Markov Models, for example, have been leveraged for this purpose with some success.

On average, different individuals of the human population share about 99.9% of their genes. In other words, approximately only one out of every thousand bases would differ between any two person. The high error rates involved with third generation sequencing are inevitably problematic for the purpose of characterizing individual differences that exist between members of the same species.

Genome Assembly

Genome assembly is the reconstruction of whole genome DNA sequences. This is generally done with two fundamentally different approaches.

Reference Alignment

When a reference genome is available, as one is in the case of human, newly sequenced reads could simply be aligned to the reference genome in order to characterize its properties. Such reference based assembly is quick and easy but has the disadvantage of "hiding" novel sequences and large copy number variants. In addition, reference genomes do not yet exist for most organisms.

De Novo Assembly

De novo assembly is the alternative genome assembly approach to reference alignment. It refers to the reconstruction of whole genome sequences entirely from raw sequence reads. This method would be chosen when there is no reference genome, when the species of the given organism is unknown as in metagenomics, or when there exist genetic variants of interest that may not be detected by reference genome alignment.

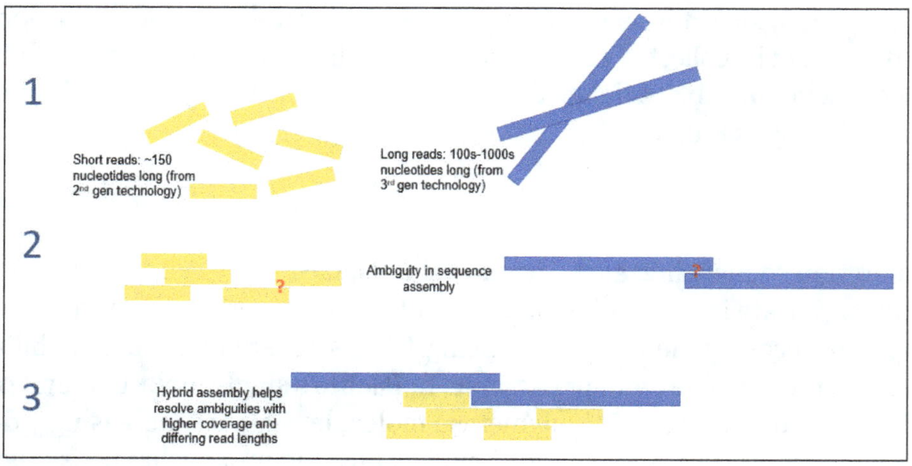

Hybrid assembly - the use of reads from 3rd gen sequencing platforms with shorts reads from 2nd gen platforms - may be used to resolve ambiguities that exist in genomes previously assembled using second generation sequencing. Short second generation reads have also been used to correct errors that exist in the long third generation reads.

Given the short reads produced by the current generation of sequencing technologies, de novo assembly is a major computational problem. It is normally approached by an iterative process of finding and connecting sequence reads with sensible overlaps. Various computational and statistical techniques, such as de bruijn graphs and overlap layout consensus graphs, have been leveraged to solve this problem. Nonetheless, due to the highly repetitive nature of eukaryotic genomes, accurate and complete reconstruction of genome sequences in de novo assembly remains challenging. Pair end reads have been posed as a possible solution, though exact fragment lengths are often unknown and must be approximated.

Hybrid Assembly

Long read lengths offered by third generation sequencing may alleviate many of the challenges currently faced by de novo genome assemblies. For example, if an entire repetitive region can be sequenced unambiguously in a single read, no computation inference would be required. Computational methods have been proposed to alleviate the issue of high error rates. For example, in one study, it was demonstrated that de novo assembly of a microbial genome using PacBio sequencing alone performed superior to that of second generation sequencing.

Third generation sequencing may also be used in conjunction with second generation sequencing. This approach is often referred to as hybrid sequencing. For example, long reads from third generation sequencing may be used to resolve ambiguities that exist in genomes previously assembled using second generation sequencing. On the other hand, short second generation reads have been used to correct errors in that exist in the long third generation reads. In general, this hybrid approach has been shown to improve de novo genome assemblies significantly.

Epigenetic Markers

DNA methylation (DNAm) – the covalent modification of DNA at CpG sites resulting in attached methyl groups – is the best understood component of epigenetic machinery. DNA modifications and resulting gene expression can vary across cell types, temporal development, with genetic ancestry, can change due to environmental stimuli and are heritable. After the discovery of DNAm, researchers have also found its correlation to diseases like cancer and autism. In this disease etiology context DNAm is an important avenue of further research.

Advantages

The current most common methods for examining methylation state require an assay that fragments DNA before standard second generation sequencing on the Illumina platform. As a result of short read length, information regarding the longer patterns of methylation are lost. Third generation sequencing technologies offer the capability for single molecule real-time sequencing of longer reads, and detection of DNA modification without the aforementioned assay.

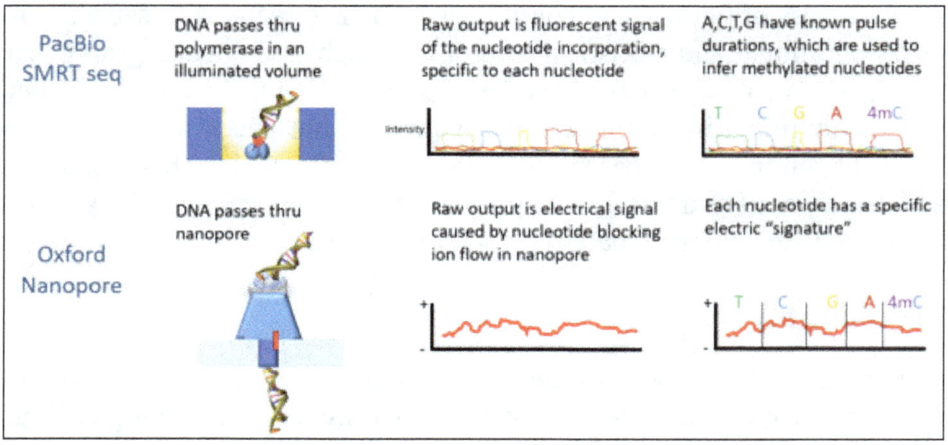

PacBio SMRT technology and Oxford Nanopore can use unaltered DNA to detect methylation.

Oxford Nanopore Technologies' MinION has been used to detect DNAm. As each DNA strand passes through a pore, it produces electrical signals which have been found to be sensitive to epigenetic changes in the nucleotides, and a hidden Markov model (HMM) was used to analyze MinION data to detect 5-methylcytosine (5mC) DNA modification. The model was trained using synthetically methylated *E. coli* DNA and the resulting signals measured by the nanopore technology. Then the trained model was used to detect 5mC in MinION genomic reads from a human cell line which already had a reference methylome. The classifier has 82% accuracy in randomly sampled singleton sites, which increases to 95% when more stringent thresholds are applied.

Other methods address different types of DNA modifications using the MinION platform. Stoiber et al. examined 4-methylcytosine (4mC) and 6-methyladenine (6mA), along with 5mC, and also created a software to directly visualize the raw MinION data in human-friendly way. Here they found that in *E. coli*, which has a known methylome, event windows of 5 base pairs long can be used to divide and statistically analyze the raw MinION electrical signals. A straightforward Mann-Whitney U test can detect modified portions of the *E. coli* sequence, as well as further split the modifications into 4mC, 6mA or 5mC regions.

It seems likely that in the future, MinION raw data will be used to detect many different epigenetic marks in DNA. PacBio sequencing has also been used to detect DNA methylation. In this platform the pulse width - the width of a fluorescent light pulse - corresponds to a specific base. In 2010 it was shown that the interpulse distance in control and methylated samples are different, and there is a "signature" pulse width for each methylation type. In 2012 using the PacBio platform the binding sites of DNA methyltransferases were characterized. The detection of N6-methylation in C Elegans was shown in 2015. DNA methylation on N^6-adenine using the PacBio platform in mouse embryonic stem cells was shown in 2016. Other forms of DNA modifications – from heavy metals, oxidation, or UV damage – are also possible avenues of research using Oxford Nanopore and PacBio third generation sequencing.

Drawbacks

Processing of the raw data – such as normalization to the median signal – was needed on MinION raw data, reducing real-time capability of the technology. Consistency of the electrical signals is still an issue, making it difficult to accurately call a nucleotide. MinION has low throughput; since multiple overlapping reads are hard to obtain, this further leads to accuracy problems of downstream DNA modification detection. Both the hidden Markov model and statistical methods used with MinION raw data require repeated observations of DNA modifications for detection, meaning that individual modified nucleotides need to be consistently present in multiple copies of the genome, e.g. in multiple cells or plasmids in the sample. For the PacBio platform, too, depending on what methylation you expect to find, coverage needs can vary. As of March 2017, other epigenetic factors like histone modifications have not been discoverable using third-generation technologies. Longer patterns of methylation are often lost because smaller contigs still need to be assembled.

Transcriptomics

Transcriptomics is the study of the transcriptome, usually by characterizing the relative abundances of messenger RNA molecules the tissue under study. According to the central dogma of molecular biology, genetic information flows from double stranded DNA molecules to single stranded mRNA molecules where they can be readily translated into function protein molecules. By studying the transcriptome, one can gain valuable insight into the regulation of gene expressions.

While expression levels as the gene level can be more or less accurately depicted by second generation sequencing, transcript level information is still an important challenge. As a consequence, the role of alternative splicing in molecular biology remains largely elusive. Third generation sequencing technologies hold promising prospects in resolving this issue by enabling sequencing of mRNA molecules at their full lengths.

Alternative Splicing

Alternative splicing (AS) is the process by which a single gene may give rise to multiple distinct mRNA transcripts and consequently different protein translations. Some evidence suggests that AS is a ubiquitous phenomenon and may play a key role in determining the phenotypes of organisms, especially in complex eukaryotes; all eukaryotes contain genes consisting of introns that may undergo AS. In particular, it has been estimated that AS occurs in 95% of all human multi-exon genes. AS has undeniable potential to influence myriad biological processes. Advancing knowledge in this area has critical implications for the study of biology in general.

Transcript Reconstruction

The current generation of sequencing technologies produce only short reads, putting tremendous limitation on the ability to detect distinct transcripts; short reads must be reverse engineered into original transcripts that could have given rise to the resulting read observations. This task is further complicated by the highly variable expression levels across transcripts, and consequently variable read coverages across the sequence of the gene. In addition, exons may be shared among individual transcripts, rendering unambiguous inferences essentially impossible. Existing computational methods make inferences based on the accumulation of short reads at various sequence locations often by making simplifying assumptions. Cufflinks takes a parsimonious approach, seeking to explain all the reads with the fewest possible number of transcripts. On the other hand, StringTie attempts to simultaneously estimate transcript abundances while assembling the reads. These methods, while reasonable, may not always identify real transcripts.

A study surveyed 25 different existing transcript reconstruction protocols. Its evidence suggested that existing methods are generally weak in assembling transcripts, though the ability to detect individual exons are relatively intact. According to the estimates, average sensitivity to detect exons across the 25 protocols is 80% for *Caenorhabditis elegans* genes. In comparison, transcript identification sensitivity decreases to 65%. For human, the study reported an exon detection sensitivity averaging to 69% and transcript detection sensitivity had an average of mere 33%. In other words, for human, existing methods are able to identify less than half of all existing transcript.

Third generation sequencing technologies have demonstrated promising prospects in solving the problem of transcript detection as well as mRNA abundance estimation at the level of transcripts. While error rates remain high, third generation sequencing technologies have the capability to produce much longer read lengths. Pacific Bioscience has introduced the iso-seq platform, proposing to sequence mRNA molecules at their full lengths. It is anticipated that Oxford Nanopore will put forth similar technologies. The trouble with higher error rates may be alleviated by supplementary high quality short reads. This approach has been previously tested and reported to reduce the error rate by more than 3 folds.

Metagenomics

Metagenomics is the analysis of genetic material recovered directly from environmental samples.

Advantages

The main advantage for third-generation sequencing technologies in metagenomics is their speed of sequencing in comparison to second generation techniques. Speed of sequencing is important for example in the clinical setting (i.e. pathogen identification), to allow for efficient diagnosis and timely clinical actions.

Oxford Nanopore's MinION was used in 2015 for real-time metagenomic detection of pathogens in complex, high-background clinical samples. The first Ebola virus (EBV) read was sequenced 44 seconds after data acquisition. There was uniform mapping of reads to genome; at least one read mapped to >88% of the genome. The relatively long reads allowed for sequencing of a near-complete viral genome to high accuracy (97–99% identity) directly from a primary clinical sample.

A common phylogenetic marker for microbial community diversity studies is the 16S ribosomal RNA gene. Both MinION and PacBio's SMRT platform have been used to sequence this gene. In this context the PacBio error rate was comparable to that of shorter reads from 454 and Illumina's MiSeq sequencing platforms.

Drawbacks

MinION's high error rate (~10-40%) prevented identification of antimicrobial resistance markers, for which single nucleotide resolution is necessary. For the same reason, eukaryotic pathogens were not identified. Ease of carryover contamination when re-using the same flow cell (standard wash protocols don't work) is also a concern. Unique barcodes may allow for more multiplexing. Furthermore, performing accurate species identification for bacteria, fungi and parasites is very difficult, as they share a larger portion of the genome, and some only differ by <5%.

The per base sequencing cost is still significantly more than that of MiSeq. However, the prospect of supplementing reference databases with full-length sequences from organisms below the limit of detection from the Sanger approach; this could possibly greatly help the identification of organisms in metagenomics.

APPLICATIONS OF NEXT-GENERATION SEQUENCING

Next generation sequencing has enabled researchers to collect vast quantities of genomic sequencing data. This technology has a plethora of applications, such as: diagnosing and understanding complex diseases; whole-genome sequencing; analysis of epigenetic modifications; mitochondrial sequencing; transcriptome sequencing – understanding how altered expression of genetic variants affects an organism; and exome sequencing – mutations in the exome are thought to contain up to 90% of mutations in the human genome, which leads to disease. DNA techniques have been used to identify and isolate genes responsible for certain diseases, and provide the correct copy of the defective gene known as 'gene therapy'.

A large focus area in gene therapy is cancer treatment – one potential method would be to introduce an antisense RNA (which specifically prevents the synthesis of a targeted protein) to the oncogene, which is triggered to form tumorous cells. Another method is named 'suicide gene therapy' which introduces genes to kill cancer cells selectively. Many genetic codes for toxic proteins and enzymes are known, and introduction of these genes into tumor cells would result in cell death. The difficulty in this method is to ensure a very precise delivery system to prevent killing healthy cells.

These methods are made possible by sequencing to analyze tumor genomes, allowing medical experts to tailor chemotherapy and other cancer treatments more effectively to their patients' unique genetic composition, revolutionizing the diagnostic stages of personalized medicine.

As the cost of DNA sequencing goes down, it will become more widespread, which brings a number of issues. Sequencing produces huge volumes of data, and there are many computational challenges associated with processing and storing the data. There are also ethical issues, such as the ownership of an individual's DNA when the DNA is sequenced. DNA sequencing data must be stored securely, since there are concerns that insurance groups, mortgage brokers and employers may use this data to modify insurance quotes or distinguish between candidates. Sequencing may also help to find out whether an individual has an increased risk to a particular disease, but whether the patient is informed or if there is a cure for the disease is another issue altogether.

References

- Dna-sequence-data-analysis-starting-off-in-bioinformatics: towardsdatascience.com, Retrieved 13 April, 2020
- Mabroukeh, N. R.; Ezeife, C. I. (2010). "A taxonomy of sequential pattern mining algorithms". ACM Computing Surveys. 43: 1–41. CiteSeerX 10.1.1.332.4745. doi:10.1145/1824795.1824798
- Human-Genome-Project, event: britannica.com, Retrieved 23 August, 2020
- Mount DM. (2004). Bioinformatics: Sequence and Genome Analysis (2nd ed.). Cold Spring Harbor Laboratory Press: Cold Spring Harbor, NY. ISBN 978-0-87969-608-5
- Next-generation-sequencing, Third-generation-sequencing: atdbio.com, Retrieved 26 April, 2020

Phylogenetic Analysis

3 CHAPTER

The study of evolutionary history and their relationships among individuals and group of organisms is called phylogenetics. Such estimation of evolutionary relationships among biological species is done through an analysis known as phylogenetic analysis. The topics elaborated in this chapter will help in gaining a better perspective about phylogenetic analysis.

PHYLOGENETICS

In Biology, Phylogenetics is the study of the ancestral relatedness of groups of organisms, whether alive or extinct. Classification of the natural world into meaningful and useful categories has long been a basic human impulse and is systematically evident at least since time of ancient Greece. Dominant for close to 2,000 years in the West was the notion of a "Great Chain of Being," or scala naturae, which emphasized a static notion of reality and depicted a hierarchy that started from matter and nature (such as rocks) and moved upward to humans, angels, and eventually and highest of all, God. Since Carolus Linnaeus, and especially since Charles Darwin, classificatory schemes have not only quickly put aside such notions as the scala naturae but also slowly moved away from postulating relationships between species based on either presumed essential traits or based on general physical similarity. The field of phylogenetics takes a functional and more scientific turn in its attempts to construct an objective depiction of evolutionary relationships between organisms based on genetic, molecular, archaeological, and historical studies and with the specific purpose of explaining, predicting, and testing similarities and differences between organisms.

Tree of Life

Ancestral relationships among species are commonly represented as phylogenetic trees (also called cladograms or dendrograms). A cladogram is a physical diagrammatic representation of a hypothesis of inferred relationship between species. The evolutionary relatedness between species is reflected by branching pattern of the tree and the relative distance between species on the tree. Any two species will have a common ancestor at the point where their respective branches are traced back to intersect. The more recent a common ancestor to humans, the more closely related it is to humans. For example, the common ancestor of modern humans and Neanderthals would be traced back to approximately 500,000 years ago. Similarly, the common ancestor of humans/Neanderthals and chimpanzees existed approximately 6.5 million years ago. Thus, since humans and Neanderthals share a common ancestor not shared by chimpanzees, we have reason to believe that humans are more closely related to Neanderthals than to chimpanzees. Similar retracing can be done farther and farther back on the branches of the tree of life to connect the common ancestry humans have with unicellular species some billions of years ago.

Estimating Relatedness

Cladograms can be constructed with the aid of technologies that estimate molecular divergences in key sequences of DNA or protein amino acids. Similar to the progress seen in estimating the age of organic substances with the use of radioactive decay technologies and carbon dating, the advent of molecular biological technologies in the later half of the 20th century have increasingly allowed scientists to more accurately estimate the degree of evolutionary relatedness at the genetic level. Taking two homologous DNA sequences in different species, one can estimate evolutionary distance by measuring the number of nucleotide substitutions that have occurred over time. Alternatively, using protein products of DNA expression, one can measure the number of amino acid substitutions that have occurred between homologous protein sequences.

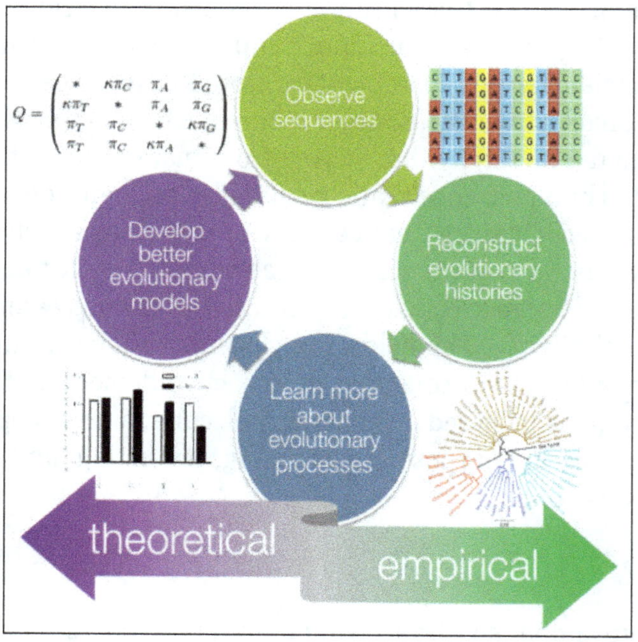

COMPUTATIONAL PHYLOGENETICS

Computational phylogenetics is the application of computational algorithms, methods, and programs to phylogenetic analyses. The goal is to assemble a phylogenetic tree representing a hypothesis about the evolutionary ancestry of a set of genes, species, or other taxa. For example, these techniques have been used to explore the family tree of hominid species and the relationships between specific genes shared by many types of organisms. Traditional phylogenetics relies on morphological data obtained by measuring and quantifying the phenotypic properties of representative organisms, while the more recent field of molecular phylogenetics uses nucleotide sequences encoding genes or amino acid sequences encoding proteins as the basis for classification. Many forms of molecular phylogenetics are closely related to and make extensive use of sequence alignment in constructing and refining phylogenetic trees, which are used to classify the evolutionary relationships between homologous genes represented in the genomes of divergent species. The phylogenetic trees constructed by computational methods are unlikely to perfectly reproduce the evolutionary tree that represents the historical relationships between the species being analyzed. The historical species tree may also differ from the historical tree of an individual homologous gene shared by those species.

Types of Phylogenetic Trees and Networks

Phylogenetic trees generated by computational phylogenetics can be either rooted or unrooted depending on the input data and the algorithm used. A rooted tree is a directed graph that explicitly identifies a most recent common ancestor (MRCA), usually an imputed sequence that is not represented in the input. Genetic distance measures can be used to plot a tree with the input sequences as leaf nodes and their distances from the root proportional to their genetic distance from the hypothesized MRCA. Identification of a root usually requires the inclusion in the input data of at least one "outgroup" known to be only distantly related to the sequences of interest.

By contrast, unrooted trees plot the distances and relationships between input sequences without making assumptions regarding their descent. An unrooted tree can always be produced from a rooted tree, but a root cannot usually be placed on an unrooted tree without additional data on divergence rates, such as the assumption of the molecular clock hypothesis.

The set of all possible phylogenetic trees for a given group of input sequences can be conceptualized as a discretely defined multidimensional "tree space" through which search paths can be traced by optimization algorithms. Although counting the total number of trees for a nontrivial number of input sequences can be complicated by variations in the definition of a tree topology, it is always true that there are more rooted than unrooted trees for a given number of inputs and choice of parameters.

Both rooted and unrooted phylogenetic trees can be further generalized to rooted or unrooted phylogenetic networks, which allow for the modeling of evolutionary phenomena such as hybridization or horizontal gene transfer.

Coding Characters and Defining Homology

Morphological Analysis

The basic problem in morphological phylogenetics is the assembly of a matrix representing a mapping from each of the taxa being compared to representative measurements for each of the phenotypic characteristics being used as a classifier. The types of phenotypic data used to construct this matrix depend on the taxa being compared; for individual species, they may involve measurements of average body size, lengths or sizes of particular bones or other physical features, or even behavioral manifestations. Of course, since not every possible phenotypic characteristic could be measured and encoded for analysis, the selection of which features to measure is a major inherent obstacle to the method. The decision of which traits to use as a basis for the matrix necessarily represents a hypothesis about which traits of a species or higher taxon are evolutionarily relevant. Morphological studies can be confounded by examples of convergent evolution of phenotypes. A major challenge in constructing useful classes is the high likelihood of inter-taxon overlap in the distribution of the phenotype's variation. The inclusion of extinct taxa in morphological analysis is often difficult due to absence of or incomplete fossil records, but has been shown to have a significant effect on the trees produced; in one study only the inclusion of extinct species of apes produced a morphologically derived tree that was consistent with that produced from molecular data.

Some phenotypic classifications, particularly those used when analyzing very diverse groups of taxa, are discrete and unambiguous; classifying organisms as possessing or lacking a tail, for example, is straightforward in the majority of cases, as is counting features such as eyes or vertebrae. However, the most appropriate representation of continuously varying phenotypic measurements is a controversial problem without a general solution. A common method is simply to sort the measurements of interest into two or more classes, rendering continuous observed variation as discretely classifiable (e.g., all examples with humerus bones longer than a given cutoff are scored as members of one state, and all members whose humerus bones are shorter than the cutoff are scored as members of a second state). This results in an easily manipulated data set but has been criticized for poor reporting of the basis for the class definitions and for sacrificing information compared to methods that use a continuous weighted distribution of measurements.

Because morphological data is extremely labor-intensive to collect, whether from literature sources or from field observations, reuse of previously compiled data matrices is not uncommon, although this may propagate flaws in the original matrix into multiple derivative analyses.

Molecular Analysis

The problem of character coding is very different in molecular analyses, as the characters in biological sequence data are immediate and discretely defined - Distinct nucleotides in DNA or RNA sequences and distinct amino acids in protein sequences. However, defining homology can be challenging due to the inherent difficulties of multiple sequence alignment. For a given gapped MSA, several rooted phylogenetic trees can be constructed that vary in their interpretations of which changes are "mutations" versus ancestral characters, and which events are insertion mutations or deletion mutations. For example, given only a pairwise alignment with a gap region, it is impossible to determine whether one sequence bears an insertion mutation or the other carries a deletion. The problem is magnified in MSAs with unaligned and nonoverlapping gaps. In practice, sizable regions of a calculated alignment may be discounted in phylogenetic tree construction to avoid integrating noisy data into the tree calculation.

Distance-matrix Methods

Distance-matrix methods of phylogenetic analysis explicitly rely on a measure of "genetic distance" between the sequences being classified, and therefore they require an MSA as an input. Distance is often defined as the fraction of mismatches at aligned positions, with gaps either ignored or counted as mismatches. Distance methods attempt to construct an all-to-all matrix from the sequence query set describing the distance between each sequence pair. From this is constructed a phylogenetic tree that places closely related sequences under the same interior node and whose branch lengths closely reproduce the observed distances between sequences. Distance-matrix methods may produce either rooted or unrooted trees, depending on the algorithm used to calculate them. They are frequently used as the basis for progressive and iterative types of multiple sequence alignments. The main disadvantage of distance-matrix methods is their inability to efficiently use information about local high-variation regions that appear across multiple subtrees.

UPGMA and WPGMA

The UPGMA (Unweighted Pair Group Method with Arithmetic mean) and WPGMA (Weighted Pair Group Method with Arithmetic mean) methods produce rooted trees and require a constant-rate assumption - that is, it assumes an ultrametric tree in which the distances from the root to every branch tip are equal.

Neighbor-joining

Neighbor-joining methods apply general cluster analysis techniques to sequence analysis using genetic distance as a clustering metric. The simple neighbor-joining method produces unrooted trees, but it does not assume a constant rate of evolution (i.e., a molecular clock) across lineages.

Fitch–Margoliash Method

The Fitch–Margoliash method uses a weighted least squares method for clustering based on genetic distance. Closely related sequences are given more weight in the tree construction process to correct for the increased inaccuracy in measuring distances between distantly related sequences. The distances used as input to the algorithm must be normalized to prevent large artifacts in computing relationships between closely related and distantly related groups. The distances calculated by this method must be linear; the linearity criterion for distances requires that the expected values of the branch lengths for two individual branches must equal the expected value of the sum of the two branch distances - a property that applies to biological sequences only when they have been corrected for the possibility of back mutations at individual sites. This correction is done through the use of a substitution matrix such as that derived from the Jukes-Cantor model of DNA evolution. The distance correction is only necessary in practice when the evolution rates differ among branches. Another modification of the algorithm can be helpful, especially in case of concentrated distances (please report to concentration of measure phenomenon and curse of dimensionality): that modification, described in, has been shown to improve the efficiency of the algorithm and its robustness.

The least-squares criterion applied to these distances is more accurate but less efficient than the neighbor-joining methods. An additional improvement that corrects for correlations between distances that arise from many closely related sequences in the data set can also be applied at increased computational cost. Finding the optimal least-squares tree with any correction factor is NP-complete, so heuristic search methods like those used in maximum-parsimony analysis are applied to the search through tree space.

Using Outgroups

Independent information about the relationship between sequences or groups can be used to help reduce the tree search space and root unrooted trees. Standard usage of distance-matrix methods involves the inclusion of at least one outgroup sequence known to be only distantly related to the sequences of interest in the query set. This usage can be seen as a type of experimental control. If the outgroup has been appropriately chosen, it will have a much greater genetic distance and thus a longer branch length than any other sequence, and it will appear near the root of a rooted tree. Choosing an appropriate outgroup requires the selection of a sequence that is moderately related to the sequences of interest; too close a relationship defeats the purpose of the outgroup and too distant adds noise to the analysis. Care should also be taken to avoid situations in which the species from which the sequences were taken are distantly related, but the gene encoded by the sequences is highly conserved across lineages. Horizontal gene transfer, especially between otherwise divergent bacteria, can also confound outgroup usage.

Maximum Parsimony

Maximum parsimony (MP) is a method of identifying the potential phylogenetic tree that requires the smallest total number of evolutionary events to explain the observed sequence data. Some ways of scoring trees also include a "cost" associated with particular types of evolutionary events and attempt to locate the tree with the smallest total cost. This is a useful approach in cases where not every possible type of event is equally likely - for example, when particular nucleotides or amino acids are known to be more mutable than others.

The most naive way of identifying the most parsimonious tree is simple enumeration - considering each possible tree in succession and searching for the tree with the smallest score. However, this is only possible for a relatively small number of sequences or species because the problem of identifying the most parsimonious tree is known to be NP-hard; consequently a number of heuristic search methods for optimization have been developed to locate a highly parsimonious tree, if not the best in the set. Most such methods involve a steepest descent-style minimization mechanism operating on a tree rearrangement criterion.

Branch and Bound

The branch and bound algorithm is a general method used to increase the efficiency of searches for near-optimal solutions of NP-hard problems first applied to phylogenetics in the early 1980s. Branch and bound is particularly well suited to phylogenetic tree construction because it inherently requires dividing a problem into a tree structure as it subdivides the problem space into smaller regions. As its name implies, it requires as input both a branching rule (in the case of phylogenetics, the addition of the next species or sequence to the tree) and a bound (a rule that excludes certain regions of the search space from consideration, thereby assuming that the optimal solution cannot occupy that region). Identifying a good bound is the most challenging aspect of the algorithm's application to phylogenetics. A simple way of defining the bound is a maximum number of assumed evolutionary changes allowed per tree. A set of criteria known as Zharkikh's rules severely limit the search space by defining characteristics shared by all candidate "most parsimonious" trees. The two most basic rules require the elimination of all but one redundant sequence (for cases where multiple observations have produced identical data) and the elimination of character sites at which two or more states do not occur in at least two species. Under ideal conditions these rules and their associated algorithm would completely define a tree.

Sankoff-Morel-Cedergren Algorithm

The Sankoff-Morel-Cedergren algorithm was among the first published methods to simultaneously produce an MSA and a phylogenetic tree for nucleotide sequences. The method uses a maximum parsimony calculation in conjunction with a scoring function that penalizes gaps and mismatches, thereby favoring the tree that introduces a minimal number of such events (an alternative view holds that the trees to be favored are those that maximize the amount of sequence similarity that can be interpreted as homology, a point of view that may lead to different optimal trees). The imputed sequences at the interior nodes of the tree are scored and summed over all the nodes in each possible tree. The lowest-scoring tree sum provides both an optimal tree and an optimal MSA given the scoring function. Because the method is highly computationally intensive, an approximate method in which initial guesses for the interior alignments are refined one node at a time. Both the full and the approximate version are in practice calculated by dynamic programming.

MALIGN and POY

More recent phylogenetic tree/MSA methods use heuristics to isolate high-scoring, but not necessarily optimal, trees. The MALIGN method uses a maximum-parsimony technique to compute a multiple alignment by maximizing a cladogram score, and its companion POY uses an iterative method that couples the optimization of the phylogenetic tree with improvements in the corresponding MSA. However, the use of these methods in constructing evolutionary hypotheses has been criticized as biased due to the deliberate construction of trees reflecting minimal evolutionary events. This, in turn, has been countered by the view that such methods should be seen as heuristic approaches to find the trees that maximize the amount of sequence similarity that can be interpreted as homology.

Maximum Likelihood

The maximum likelihood method uses standard statistical techniques for inferring probability distributions to assign probabilities to particular possible phylogenetic trees. The method requires a substitution model to assess the probability of particular mutations; roughly, a tree that requires more mutations at interior nodes to explain the observed phylogeny will be assessed as having a lower probability. This is broadly similar to the maximum-parsimony method, but maximum likelihood allows additional statistical flexibility by permitting varying rates of evolution across both lineages and sites. In fact, the method requires that evolution at different sites and along different lineages must be statistically independent. Maximum likelihood is thus well suited to the analysis of distantly related sequences, but it is believed to be computationally intractable to compute due to its NP-hardness.

The "pruning" algorithm, a variant of dynamic programming, is often used to reduce the search space by efficiently calculating the likelihood of subtrees. The method calculates the likelihood for each site in a "linear" manner, starting at a node whose only descendants are leaves (that is, the tips of the tree) and working backwards toward the "bottom" node in nested sets. However, the trees produced by the method are only rooted if the substitution model is irreversible, which is not generally true of biological systems. The search for the maximum-likelihood tree also includes a branch length optimization component that is difficult to improve upon algorithmically; general global optimization tools such as the Newton-Raphson method are often used.

Some tools that use maximum likelihood to infer phylogenetic trees from variant allelic frequency data (VAFs) include AncesTree and CITUP.

Bayesian Inference

Bayesian inference can be used to produce phylogenetic trees in a manner closely related to the maximum likelihood methods. Bayesian methods assume a prior probability distribution of the possible trees, which may simply be the probability of any one tree among all the possible trees that could be generated from the data, or may be a more sophisticated estimate derived from the assumption that divergence events such as speciation occur as stochastic processes. The choice of prior distribution is a point of contention among users of Bayesian-inference phylogenetics methods.

Implementations of Bayesian methods generally use Markov chain Monte Carlo sampling algorithms, although the choice of move set varies; selections used in Bayesian phylogenetics include circularly permuting leaf nodes of a proposed tree at each step and swapping descendant subtrees of a random internal node between two related trees. The use of Bayesian methods in phylogenetics has been controversial, largely due to incomplete specification of the choice of move set, acceptance criterion, and prior distribution in published work. Bayesian methods are generally held to be superior to parsimony-based methods; they can be more prone to long-branch attraction than maximum likelihood techniques, although they are better able to accommodate missing data.

Whereas likelihood methods find the tree that maximizes the probability of the data, a Bayesian approach recovers a tree that represents the most likely clades, by drawing on the posterior distribution. However, estimates of the posterior probability of clades (measuring their 'support') can be quite wide of the mark, especially in clades that aren't overwhelmingly likely. As such, other methods have been put forwards to estimate posterior probability.

Some tools that use Bayesian inference to infer phylogenetic trees from variant allelic frequency data (VAFs) include Canopy, EXACT, and PhyloWGS.

Model Selection

Molecular phylogenetics methods rely on a defined substitution model that encodes a hypothesis about the relative rates of mutation at various sites along the gene or amino acid sequences being studied. At their simplest, substitution models aim to correct for differences in the rates of transitions and transversions in nucleotide sequences. The use of substitution models is necessitated by the fact that the genetic distance between two sequences increases linearly only for a short time after the two sequences diverge from each other (alternatively, the distance is linear only shortly before coalescence). The longer the amount of time after divergence, the more likely it becomes that two mutations occur at the same nucleotide site. Simple genetic distance calculations will thus undercount the number of mutation events that have occurred in evolutionary history. The extent of this undercount increases with increasing time since divergence, which can lead to the phenomenon of long branch attraction, or the misassignment of two distantly related but convergently evolving sequences as closely related. The maximum parsimony method is particularly susceptible to this problem due to its explicit search for a tree representing a minimum number of distinct evolutionary events.

Types of Models

All substitution models assign a set of weights to each possible change of state represented in the sequence. The most common model types are implicitly reversible because they assign the same weight to, for example, a G>C nucleotide mutation as to a C>G mutation. The simplest possible model, the Jukes-Cantor model, assigns an equal probability to every possible change of state for a given nucleotide base. The rate of change between any two distinct nucleotides will be one-third of the overall substitution rate. More advanced models distinguish between transitions and transversions. The most general possible time-reversible model, called the GTR model, has six mutation rate parameters. An even more generalized model known as the general 12-parameter model breaks time-reversibility, at the cost of much additional complexity in calculating genetic distances that are consistent among multiple lineages. One possible variation on this theme adjusts the rates so that overall GC content - an important measure of DNA double helix stability - varies over time.

Models may also allow for the variation of rates with positions in the input sequence. The most obvious example of such variation follows from the arrangement of nucleotides in protein-coding genes into three-base codons. If the location of the open reading frame (ORF) is known, rates of mutation can be adjusted for position of a given site within a codon, since it is known that wobble base pairing can allow for higher mutation rates in the third nucleotide of a given codon without affecting the codon's meaning in the genetic code. A less hypothesis-driven example that does not rely on ORF identification simply assigns to each site a rate randomly drawn from a predetermined distribution, often the gamma distribution or log-normal distribution. Finally, a more conservative estimate of rate variations known as the covarion method allows autocorrelated variations in rates, so that the mutation rate of a given site is correlated across sites and lineages.

Choosing the Best Model

The selection of an appropriate model is critical for the production of good phylogenetic analyses, both because underparameterized or overly restrictive models may produce aberrant behavior when their underlying assumptions are violated, and because overly complex or overparameterized models are computationally expensive and the parameters may be overfit. The most common method of model selection is the likelihood ratio test (LRT), which produces a likelihood estimate that can be interpreted as a measure of "goodness of fit" between the model and the input data. However, care must be taken in using these results, since a more complex model with more parameters will always have a higher likelihood than a simplified version of the same model, which can lead to the naive selection of models that are overly complex. For this reason model selection computer programs will choose the simplest model that is not significantly worse than more complex substitution models. A significant disadvantage of the LRT is the necessity of making a series of pairwise comparisons between models; it has been shown that the order in which the models are compared has a major effect on the one that is eventually selected.

An alternative model selection method is the Akaike information criterion (AIC), formally an estimate of the Kullback–Leibler divergence between the true model and the model being tested. It can be interpreted as a likelihood estimate with a correction factor to penalize overparameterized models. The AIC is calculated on an individual model rather than a pair, so it is independent of the order in which models are assessed. A related alternative, the Bayesian information criterion (BIC), has a similar basic interpretation but penalizes complex models more heavily.

A comprehensive step-by-step protocol on constructing phylogenetic tree, including DNA/Amino Acid contiguous sequence assembly, multiple sequence alignment, model-test (testing best-fitting substitution models) and phylogeny reconstruction using Maximum Likelihood and Bayesian Inference, is available at Nature Protocol.

A non traditional way of evaluating the phylogenetic tree is to compare it with clustering result. One can use a Multidimensional Scaling technique, so called Interpolative Joining to do dimensionality reduction to visualize the clustering result for the sequences in 3D, and then map the phylogenetic tree onto the clustering result. A better tree usually has a higher correlation with the clustering result.

Evaluating Tree Support

As with all statistical analysis, the estimation of phylogenies from character data requires an evaluation of confidence. A number of methods exist to test the amount of support for a phylogenetic tree, either by evaluating the support for each sub-tree in the phylogeny (nodal support) or evaluating whether the phylogeny is significantly different from other possible trees (alternative tree hypothesis tests).

Nodal Support

The most common method for assessing tree support is to evaluate the statistical support for each node on the tree. Typically, a node with very low support is not considered valid in further analysis, and visually may be collapsed into a polytomy to indicate that relationships within a clade are unresolved.

Consensus Tree

Many methods for assessing nodal support involve consideration of multiple phylogenies. The consensus tree summarizes the nodes that are shared among a set of trees. In a *strict consensus,* only nodes found in every tree are shown, and the rest are collapsed into an unresolved polytomy. Less conservative methods, such as the *majority-rule consensus* tree, consider nodes that are supported by a given percentage of trees under consideration (such as at least 50%).

For example, in maximum parsimony analysis, there may be many trees with the same parsimony score. A strict consensus tree would show which nodes are found in all equally parsimonious trees, and which nodes differ. Consensus trees are also used to evaluate support on phylogenies reconstructed with Bayesian inference.

Bootstrapping and Jackknifing

In statistics, the bootstrap is a method for inferring the variability of data that has an unknown distribution using pseudoreplications of the original data. For example, given a set of 100 data points, a pseudoreplicate is a data set of the same size (100 points) randomly sampled from the original data, with replacement. That is, each original data point may be represented more than once in the pseudoreplicate, or not at all. Statistical support involves evaluation of whether the original data has similar properties to a large set of pseudoreplicates.

In phylogenetics, bootstrapping is conducted using the columns of the character matrix. Each pseudoreplicate contains the same number of species (rows) and characters (columns) randomly sampled from the original matrix, with replacement. A phylogeny is reconstructed from each pseudoreplicate, with the same methods used to reconstruct the phylogeny from the original data. For each node on the phylogeny, the nodal support is the percentage of pseudoreplicates containing that node.

The statistical rigor of the bootstrap test has been empirically evaluated using viral populations with known evolutionary histories, finding that 70% bootstrap support corresponds to a 95% probability that the clade exists. However, this was tested under ideal conditions (e.g. no change in evolutionary rates, symmetric phylogenies). In practice, values above 70% are generally supported and left to the researcher or reader to evaluate confidence. Nodes with support lower than 70% are typically considered unresolved.

Jackknifing in phylogenetics is a similar procedure, except the columns of the matrix are sampled without replacement. Pseudoreplicates are generated by randomly subsampling the data—for example, a "10% jackknife" would involve randomly sampling 10% of the matrix many times to evaluate nodal support.

Posterior Probability

Reconstruction of phylogenies using Bayesian inference generates a posterior distribution of highly probable trees given the data and evolutionary model, rather than a single "best" tree. The trees in the posterior distribution generally have many different topologies. When the input data is variant allelic frequency data (VAF), the tool EXACT can compute the probabilities of trees exactly, for small, biologically relevant tree sizes, by exhaustively searching the entire tree space.

Most Bayesian inference methods utilize a Markov-chain Monte Carlo iteration, and the initial steps of this chain are not considered reliable reconstructions of the phylogeny. Trees generated early in the chain are usually discarded as burn-in. The most common method of evaluating nodal support in a Bayesian phylogenetic analysis is to calculate the percentage of trees in the posterior distribution (post-burn-in) which contain the node.

The statistical support for a node in Bayesian inference is expected to reflect the probability that a clade really exists given the data and evolutionary model. Therefore, the threshold for accepting a node as supported is generally higher than for bootstrapping.

Step Counting Methods

Bremer support counts the number of extra steps needed to contradict a clade.

Shortcomings

These measures each have their weaknesses. For example, smaller or larger clades tend to attract larger support values than mid-sized clades, simply as a result of the number of taxa in them.

Bootstrap support can provide high estimates of node support as a result of noise in the data rather than the true existence of a clade.

Limitations and Workarounds

Ultimately, there is no way to measure whether a particular phylogenetic hypothesis is accurate or not, unless the true relationships among the taxa being examined are already known (which may happen with bacteria or viruses under laboratory conditions). The best result an empirical phylogeneticist can hope to attain is a tree with branches that are well supported by the available evidence. Several potential pitfalls have been identified:

Homoplasy

Certain characters are more likely to evolve convergently than others; logically, such characters should be given less weight in the reconstruction of a tree. Weights in the form of a model of evolution can be inferred from sets of molecular data, so that maximum likelihood or Bayesian methods can be used to analyze them. For molecular sequences, this problem is exacerbated when the taxa under study have diverged substantially. As time since the divergence of two taxa increase, so does the probability of multiple substitutions on the same site, or back mutations, all of which result in homoplasies. For morphological data, unfortunately, the only objective way to determine convergence is by the construction of a tree – a somewhat circular method. Even so, weighting homoplasious characters does indeed lead to better-supported trees. Further refinement can be brought by weighting changes in one direction higher than changes in another; for instance, the presence of thoracic wings almost guarantees placement among the pterygote insects because, although wings are often lost secondarily, there is no evidence that they have been gained more than once.

Horizontal Gene Transfer

In general, organisms can inherit genes in two ways: vertical gene transfer and horizontal gene transfer. Vertical gene transfer is the passage of genes from parent to offspring, and horizontal (also called lateral) gene transfer occurs when genes jump between unrelated organisms, a common phenomenon especially in prokaryotes; a good example of this is the acquired antibiotic resistance as a result of gene exchange between various bacteria leading to multi-drug-resistant bacterial species. There have also been well-documented cases of horizontal gene transfer between eukaryotes.

Horizontal gene transfer has complicated the determination of phylogenies of organisms, and inconsistencies in phylogeny have been reported among specific groups of organisms depending on the genes used to construct evolutionary trees. The only way to determine which genes have been acquired vertically and which horizontally is to parsimoniously assume that the largest set of genes that have been inherited together have been inherited vertically; this requires analyzing a large number of genes.

Hybrids, Speciation, Introgressions and Incomplete Lineage Sorting

The basic assumption underlying the mathematical model of cladistics is a situation where species split neatly in bifurcating fashion. While such an assumption may hold on a larger scale (bar horizontal gene transfer,), speciation is often much less orderly. Research since the cladistic method was introduced has shown that hybrid speciation, once thought rare, is in fact quite common, particularly in plants. Also paraphyletic speciation is common, making the assumption of a bifurcating pattern unsuitable, leading to phylogenetic networks rather than trees. Introgression can also move genes between otherwise distinct species and sometimes even genera, complicating phylogenetic analysis based on genes. This phenomenon can contribute to "incomplete lineage sorting" and is thought to be a common phenomenon across a number of groups. In species level analysis this can be dealt with by larger sampling or better whole genome analysis. Often the problem is avoided by restricting the analysis to fewer, not closely related specimens.

Taxon Sampling

Owing to the development of advanced sequencing techniques in molecular biology, it has become feasible to gather large amounts of data (DNA or amino acid sequences) to infer phylogenetic hypotheses. For example, it is not rare to find studies with character matrices based on whole mitochondrial genomes (~16,000 nucleotides, in many animals). However, simulations have shown that it is more important to increase the number of taxa in the matrix than to increase the number of characters, because the more taxa there are, the more accurate and more robust is the resulting phylogenetic tree. This may be partly due to the breaking up of long branches.

Phylogenetic Signal

Another important factor that affects the accuracy of tree reconstruction is whether the data analyzed actually contain a useful phylogenetic signal, a term that is used generally to denote whether a character evolves slowly enough to have the same state in closely related taxa as opposed to varying randomly. Tests for phylogenetic signal exist.

Continuous Characters

Morphological characters that sample a continuum may contain phylogenetic signal, but are hard to code as discrete characters. Several methods have been used, one of which is gap coding, and there are variations on gap coding. In the original form of gap coding:group means for a character are first ordered by size. The pooled within-group standard deviation is calculated and differences between adjacent means are compared relative to this standard deviation. Any pair of adjacent means is considered different and given different integer scores if the means are separated by a "gap" greater than the within-group standard deviation times some arbitrary constant.

If more taxa are added to the analysis, the gaps between taxa may become so small that all information is lost. Generalized gap coding works around that problem by comparing individual pairs of taxa rather than considering one set that contains all of the taxa.

Missing Data

In general, the more data that are available when constructing a tree, the more accurate and reliable the resulting tree will be. Missing data are no more detrimental than simply having fewer data, although the impact is greatest when most of the missing data are in a small number of taxa. Concentrating the missing data across a small number of characters produces a more robust tree.

Role of Fossils

Because many characters involve embryological, or soft-tissue or molecular characters that (at best) hardly ever fossilize, and the interpretation of fossils is more ambiguous than that of living taxa, extinct taxa almost invariably have higher proportions of missing data than living ones. However, despite these limitations, the inclusion of fossils is invaluable, as they can provide information in sparse areas of trees, breaking up long branches and constraining intermediate character states; thus, fossil taxa contribute as much to tree resolution as modern taxa. Fossils can also constrain the age of lineages and thus demonstrate how consistent a tree is with the stratigraphic record; stratocladistics incorporates age information into data matrices for phylogenetic analyses.

PHYLOGENETIC TREE

A phylogenetic tree or evolutionary tree is a branching diagram or "tree" showing the evolutionary relationships among various biological species or other entities—their phylogeny—based upon similarities and differences in their physical or genetic characteristics. All life on Earth is part of a single phylogenetic tree, indicating common ancestry.

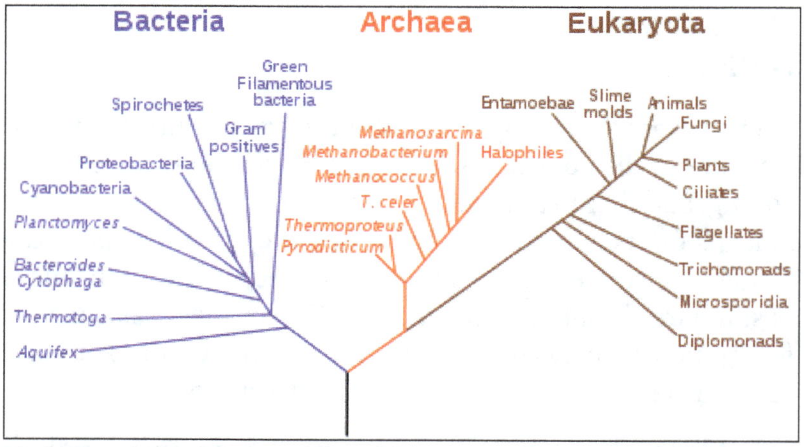

A speculatively rooted tree for rRNA genes, showing the three life domains: bacteria, archaea, and eukaryota. The black branch at the bottom of the phylogenetic tree connects the three branches of living organisms to the last universal common ancestor.

In a rooted phylogenetic tree, each node with descendants represents the inferred most recent common ancestor of those descendants, and the edge lengths in some trees may be interpreted as time estimates. Each node is called a taxonomic unit. Internal nodes are generally called hypothetical taxonomic units, as they cannot be directly observed. Trees are useful in fields of biology such as bioinformatics, systematics, and phylogenetics. Unrooted trees illustrate only the relatedness of the leaf nodes and do not require the ancestral root to be known or inferred.

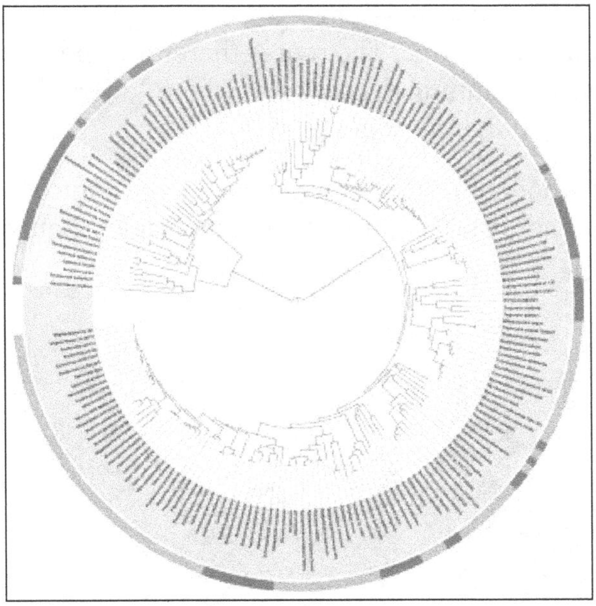

A highly resolved, automatically generated tree of life, based on completely sequenced genomes.

Properties

Rooted Tree

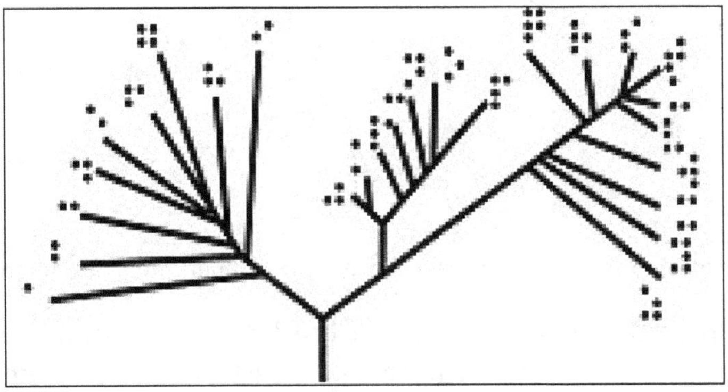

Rooted phylogenetic tree, optimized for blind people. The lowest point of the tree is the root, which symbolizes the universal common ancestor to all living beings. The tree branches out into three main groups: Bacteria (left branch, letters a to i), Archea (middle branch, letters j to p) and Eukaryota (right branch, letters q to z). Each letter corresponds to a group of organisms, listed below this description. These letters and the description should be converted to Braille font, and printed using a Braille printer. The figure can be 3D printed by copying the png file and using Cura or other software to generate the Gcode for 3D printing.

A rooted phylogenetic tree is a directed tree with a unique node — the root — corresponding to the (usually imputed) most recent common ancestor of all the entities at the leaves of the tree. The root node does not have a parent node, but serves as the parent of all other nodes in the tree. The root is therefore a node of degree 2 while other internal nodes have a minimum degree of 3 (where "degree" here refers to the total number of incoming and outgoing edges).

The most common method for rooting trees is the use of an uncontroversial outgroup—close enough to allow inference from trait data or molecular sequencing, but far enough to be a clear outgroup.

Unrooted Tree

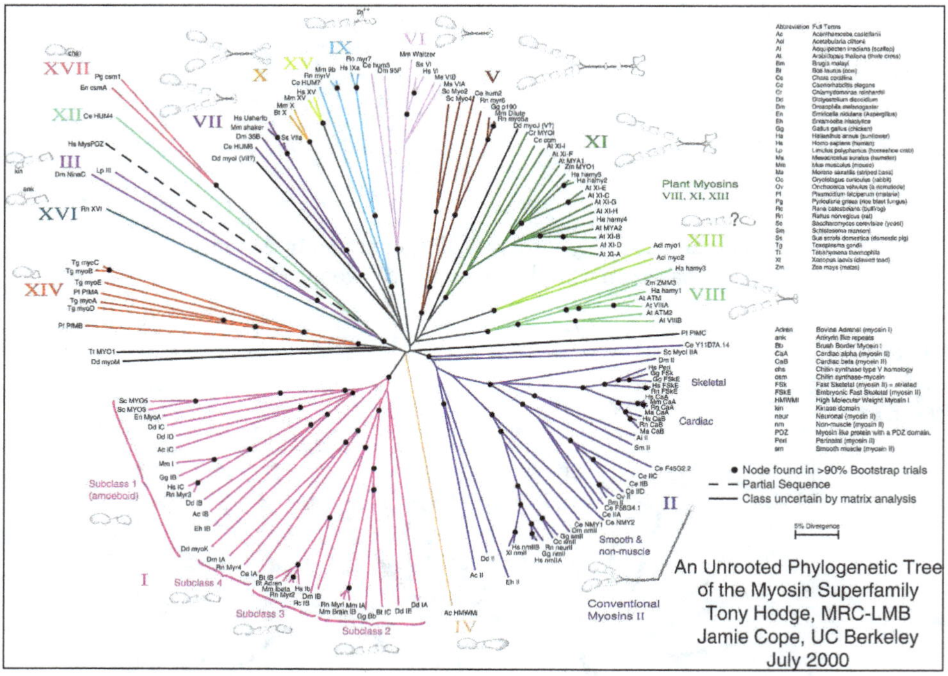

An unrooted phylogenetic tree for myosin, a superfamily of proteins.

Unrooted trees illustrate the relatedness of the leaf nodes without making assumptions about ancestry. They do not require the ancestral root to be known or inferred. Unrooted trees can always be generated from rooted ones by simply omitting the root. By contrast, inferring the root of an unrooted tree requires some means of identifying ancestry. This is normally done by including an outgroup in the input data so that the root is necessarily between the outgroup and the rest of the taxa in the tree, or by introducing additional assumptions about the relative rates of evolution on each branch, such as an application of the molecular clock hypothesis.

Bifurcating versus Multifurcating

Both rooted and unrooted trees can be either bifurcating or multifurcating. A rooted bifurcating tree has exactly two descendants arising from each interior node (that is, it forms a binary tree), and an unrooted bifurcating tree takes the form of an unrooted binary tree, a free tree with exactly three neighbors at each internal node. In contrast, a rooted multifurcating tree may have more than two children at some nodes and an unrooted multifurcating tree may have more than three neighbors at some nodes.

Labeled versus Unlabeled

Both rooted and unrooted trees can be either labeled or unlabeled. A labeled tree has specific values assigned to its leaves, while an unlabeled tree, sometimes called a tree shape, defines a topology only.

Enumerating Trees

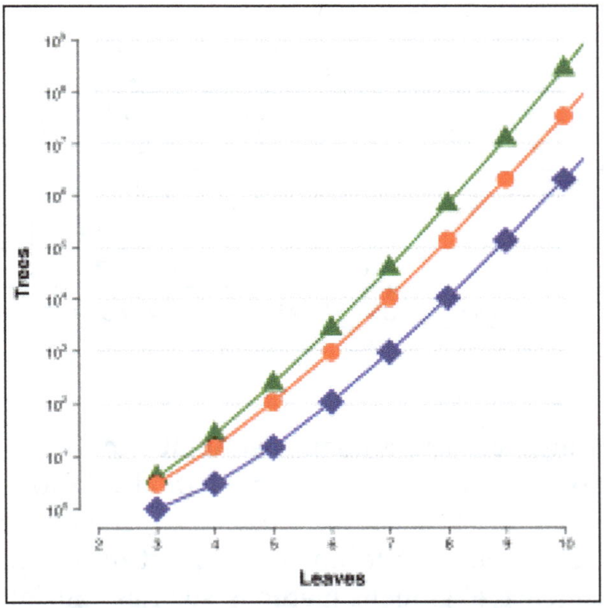

Increase in the total number of phylogenetic trees as a function of the number of labeled leaves: unrooted binary trees (blue diamonds), rooted binary trees (red circles), and rooted multifurcating or binary trees (green: triangles). The Y-axis scale is logarithmic.

The number of possible trees for a given number of leaf nodes depends on the specific type of tree, but there are always more labeled than unlabeled trees, more multifurcating than bifurcating trees, and more rooted than unrooted trees. The last distinction is the most biologically relevant; it arises because there are many places on an unrooted tree to put the root. For bifurcating labeled trees, the total number of rooted trees is:

$(2n-3)!! = \dfrac{(2n-3)!}{2^{n-2}(n-2)!}$ for $n \geq 2$, where n represents the number of leaf nodes.

For bifurcating labeled trees, the total number of unrooted trees is:

$(2n-5)!! = \dfrac{(2n-5)!}{2^{n-3}(n-3)!}$ for $n \geq 3$.

Among labeled bifurcating trees, the number of unrooted trees with n leaves is equal to the number of rooted trees with $n-1$ leaves.

The number of rooted trees grows quickly as a function of the number of tips. For 10 tips, there are more than 34×10^6 possible bifurcating trees, and the number of multifurcating trees rises faster, with ca. 7 times as many of the latter as of the former.

Counting trees				
Labeled leaves	Binary unrooted trees	Binary rooted trees	Multifurcating rooted trees	All possible rooted trees
1	1	1	0	1
2	1	1	0	1
3	1	3	1	4
4	3	15	11	26
5	15	105	131	236
6	105	945	1,807	2,752
7	945	10,395	28,813	39,208
8	10,395	135,135	524,897	660,032
9	135,135	2,027,025	10,791,887	12,818,912
10	2,027,025	34,459,425	247,678,399	282,137,824

Construction

Phylogenetic trees composed with a nontrivial number of input sequences are constructed using computational phylogenetics methods. Distance-matrix methods such as neighbor-joining or UPGMA, which calculate genetic distance from multiple sequence alignments, are simplest to implement, but do not invoke an evolutionary model. Many sequence alignment methods such as ClustalW also create trees by using the simpler algorithms (i.e. those based on distance) of tree construction. Maximum parsimony is another simple method of estimating phylogenetic trees, but implies an implicit model of evolution (i.e. parsimony). More advanced methods use the optimality criterion of maximum likelihood, often within a Bayesian framework, and apply an explicit model of evolution to phylogenetic tree estimation. Identifying the optimal tree using many of these techniques is NP-hard, so heuristic search and optimization methods are used in combination with tree-scoring functions to identify a reasonably good tree that fits the data.

Tree-building methods can be assessed on the basis of several criteria:

- Efficiency (how long does it take to compute the answer, how much memory does it need?).
- Power (does it make good use of the data, or is information being wasted?).

- Consistency (will it converge on the same answer repeatedly, if each time given different data for the same model problem?).
- Robustness (does it cope well with violations of the assumptions of the underlying model?).
- Falsifiability (does it alert us when it is not good to use, i.e. when assumptions are violated?).

Tree-building techniques have also gained the attention of mathematicians. Trees can also be built using T-theory.

Limitations of Phylogenetic Analysis

Although phylogenetic trees produced on the basis of sequenced genes or genomic data in different species can provide evolutionary insight, these analyses have important limitations. Most importantly, the trees that they generate are not necessarily correct – they do not necessarily accurately represent the evolutionary history of the included taxa. As with any scientific result, they are subject to falsification by further study (e.g., gathering of additional data, analyzing the existing data with improved methods). The data on which they are based may be noisy; the analysis can be confounded by genetic recombination, horizontal gene transfer, hybridisation between species that were not nearest neighbors on the tree before hybridisation takes place, convergent evolution, and conserved sequences.

Also, there are problems in basing an analysis on a single type of character, such as a single gene or protein or only on morphological analysis, because such trees constructed from another unrelated data source often differ from the first, and therefore great care is needed in inferring phylogenetic relationships among species. This is most true of genetic material that is subject to lateral gene transfer and recombination, where different haplotype blocks can have different histories. In these types of analysis, the output tree of a phylogenetic analysis of a single gene is an estimate of the gene's phylogeny (i.e. a gene tree) and not the phylogeny of the taxa (i.e. species tree) from which these characters were sampled, though ideally, both should be very close. For this reason, serious phylogenetic studies generally use a combination of genes that come from different genomic sources (e.g., from mitochondrial or plastid vs. nuclear genomes), or genes that would be expected to evolve under different selective regimes, so that homoplasy (false homology) would be unlikely to result from natural selection.

When extinct species are included as terminal nodes in an analysis (rather than, for example, to constrain internal nodes), they are considered not to represent direct ancestors of any extant species. Extinct species do not typically contain high-quality DNA. The range of useful DNA materials has expanded with advances in extraction and sequencing technologies. Development of technologies able to infer sequences from smaller fragments, or from spatial patterns of DNA degradation products, would further expand the range of DNA considered useful.

Phylogenetic trees can also be inferred from a range of other data types, including morphology, the presence or absence of particular types of genes, insertion and deletion events – and any other observation thought to contain an evolutionary signal. Phylogenetic networks are used when bifurcating trees are not suitable, due to these complications which suggest a more reticulate evolutionary history of the organisms sampled.

TYPES OF PHYLOGENETIC TREES

Dendrogram

A dendrogram is a network structure. It is constituted of a root node that gives birth to several nodes connected by edges or branches. The last nodes of the hierarchy are called leaves. In the following example, the CEO is the root node. He manages 2 managers that manage 8 employees (the leaves).

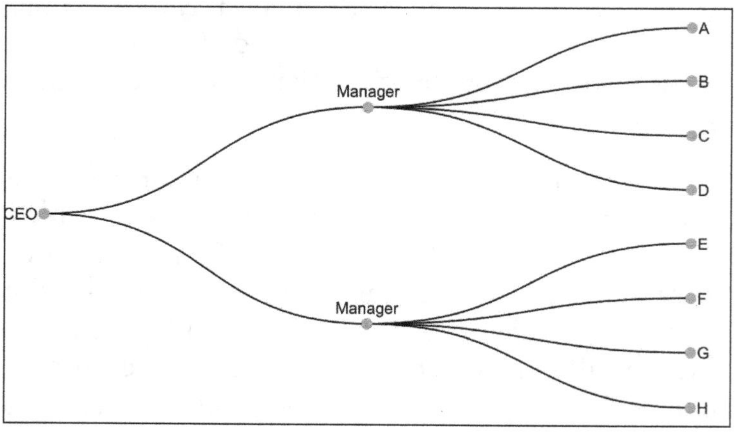

Two type of dendrogram exist, resulting from two types of dataset:

- A hierarchic dataset provides the links between nodes explicitely.
- The result of a clustering algorythm can be visualized as a dendrogram.

Dendrogram from Hierarchic Data

Dendrogram from Clustering

A distance matrix that provides the distance between all pairs of 28 major cities. Note that this kind of matrix can be computed from a multivariate dataset, computing distance between each pair of individual using correlation or euclidean distance.

	Berlin	Cairo	Caracas
Berlin	•	1795	5247
Cairo	1795	•	6338
Caracas	5247	6338	•

It is possible to perform hierarchical cluster analysis on this set of dissimilarities. Basically, this statistical method seeks to build a hierarchy of clusters: it tries to group sample that are close one from another.

The result can be seen as a dendrogram:

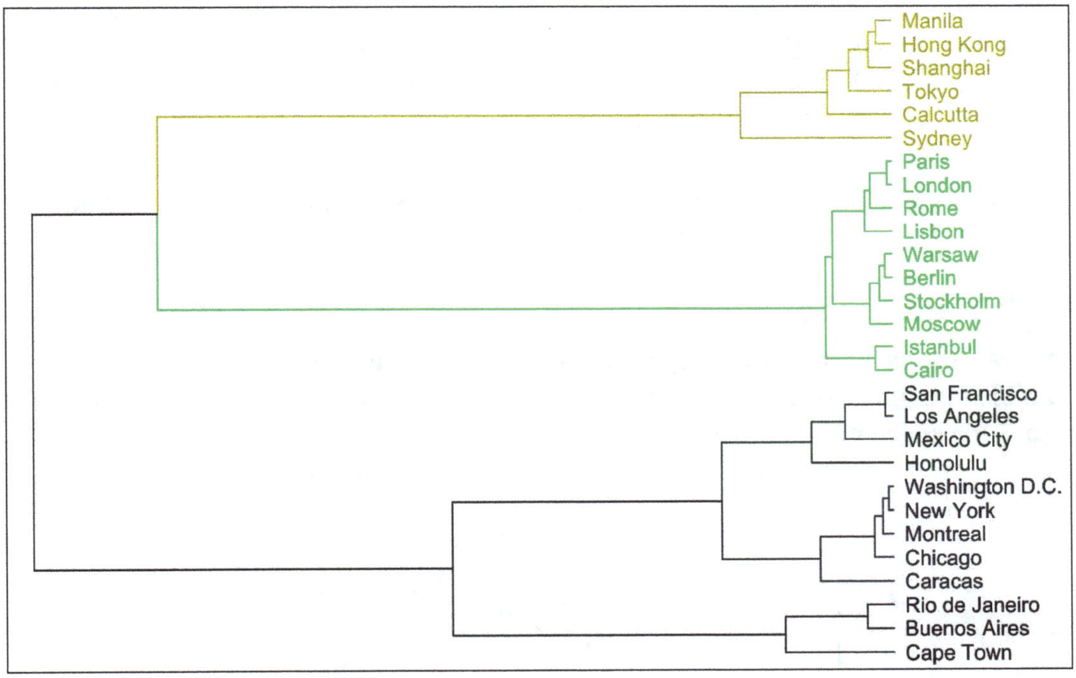

Cities that are in same geographic area tend to be clusterized together. For example, the yellow cluster is composed by all the Asian cities of the dataset. Note that the dendrogram provides even more information. For instance, Sydney appears to be a bit further to Calcutta than calcutta is from Tokyo: this can be deduce from the branch size that represents the distance.

A common task consists to compare the result of a clustering with an expected result. For instance, we can check if the countries are indeed grouped in continent using a color bar.

This graphic allows validating that the clustering indeed grouped cities by continent. There are a few discrepencies that are logical. Indeed, Mexico city has been considered as a city of South America here, altough it is probably closer from North America as suggested by the clustering.

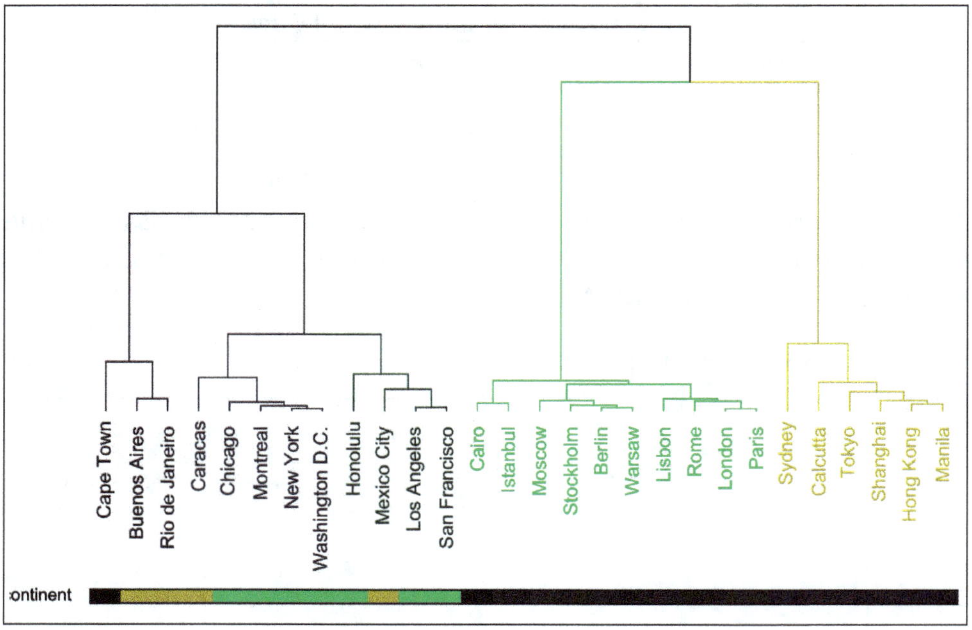

Variation

Many variations exist for dendrogram. It can be horizontal or vertical as shown before. It can also be linear or circular. The advantage of the circular verion being that it uses the graphic space more efficiently.

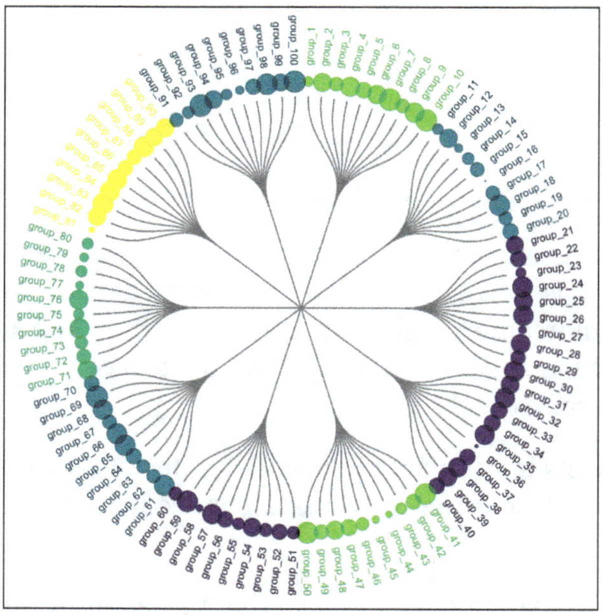

Another common variation is to display a heatmap at the bottom of the dendrogram. Indeed, it allows to visualize the distance between each sample and thus to understand why the clustering algorythm put 2 samples next to each other.

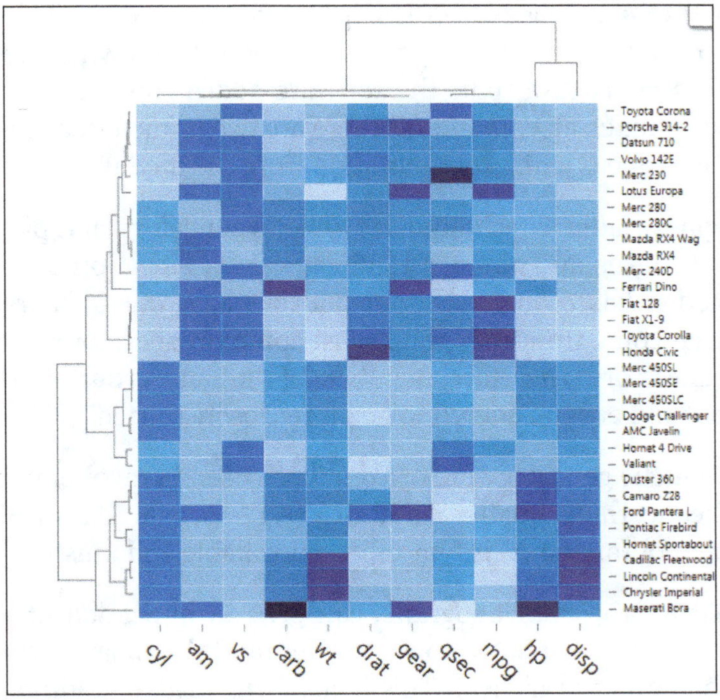

Cladogram

In genetic genealogy, a cladogram is a diagram showing genetic relationships among a group of people. It depicts their commonalities and differences. "Phylogenetic tree" is a synonym for cladogram.

The chief value of cladograms is their graphic nature. A picture is often worth many words or numbers.

One example of a cladogram is the ISOGG phylogenetic tree of the human Y chromosome. Another is a descendancy chart printed by genealogical software.

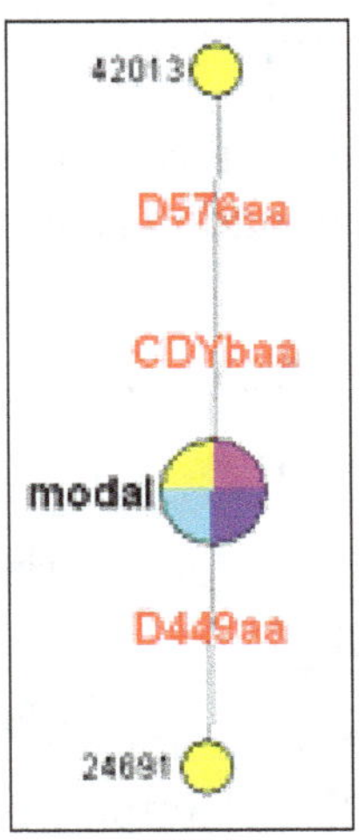

Another example is the Fluxus Network diagram to the right -- a cladogram depicting six persons' 37 Y-STR marker haplotypes for a genetic family in one surname DNA project. It consists of six taxa (plus the software-added "modal" shown in magenta; three have an exact match at 37 markers. There are two branches; one is two mutations from the modal and one is a different mutation from the modal.

A cladogram applies cladistic techniques to genetic genealogy and pictorially displays the results. "Cladistics is an approach to biological classification in which organisms are grouped together based on whether or not they have one or more shared unique characteristics that come from the group's last common ancestor and are not present in more distant ancestors. Therefore, members of the same group are thought to share a common history and are considered to be more closely related".

Cladograms for genetic genealogy differ from those used in taxonomy in focusing only on humans and concerning smaller variations (e.g., specific STR or SNP mutations). Also, data for the root (ancestor) is typically not available and must be hypothesized.

In surname Y-DNA projects, cladograms can give an indication of which branches are likely to be more closely related to each other (and thus share a common ancestor more recently than other branches). This helps focus further traditional genealogical research in that it helps identify those branches which perhaps should be working more closely together to find documentary evidence of their (more recent) connection.

The basic unit of a cladogram is a taxon (plural taxa), an individual, group or species (in short, anything the cladisticist chooses). In Y-DNA cladograms, a taxon may be an individual's haplotype or pattern of STR markers and their allele values. The taxa in a cladogram are usually inferred to be phylogenetically related. It is recommended that the cladisticist assure phylogenetic relationships before constructing a cladogram.

A frequent cladogram convention is to imply "advancement" by direction. Upward and rightward may signify greater complexity; downward and leftward, less complexity. This may not apply in genetic genealogy; a mutated haplotype is not necessarily either "better" or "worse" than the prior form.

Cladograms often rely on a parsimony criterion (e.g., Occam's razor) to infer phylogeny from molecular data. That is, the explanation requiring the fewest assumptions is most likely to be true.

- It is assumed that haplotypes with greater similarity are more closely related than those with more dissimilarity.
- And in Y-STR terms, a group's modal marker value is the assumed value of the common ancestor, i.e., the "root".

Cladograms may have a wide range of complexity, depending on the number of taxa and the closeness of relationships. For example, an exact Y-DNA match between two or more people results in a single point (taxon). As differences increase, the points become separated.

Cladogram Benefits

Cladograms have uses in both analysis of genetic data and in illustrating the interpretation of the data.

- Analysis: Cladograms isolate certain aspects by ignoring properties shared in common and highlighting differences. Also, some types of cladogram may help discover selection bias in the data.
- Illustration: Pictures are simpler for many to understand than words or numbers. A cladogram may help to illustrate key points.

Constructing a Cladogram

The steps in constructing a cladogram are:

- Gather and organize the data. This will often be Y-STR allele values by marker. The Y-STR must all be tested at the same resolution.
- Consider possible cladograms. This may involve specialized software such as Fluxus or PHYLIP. McGee Utilities can be used to format your STR data for input. (Copy and paste the McGee output into a plain-text file).
    - Fluxus may be downloaded free. An acknowledgement is required when publishing diagrams or output data. The illustration above shows a simple version of its output.
    - PHYLIP is a set of 35 programs, also free but designed for academic use. Output from "drawtree" is in the "Newick" tree format.
- Select best cladogram: The one that most closely portrays the data and illuminates the relationships.

Examples of Cladogram

Primates

The various groups of primates being compared are listed on the top. The various nodes on the diagram represent the various common ancestors between the groups. The Apes, the group containing humans, and all the common ancestors (nodes) down to the lowest ape are considered a clade, or group of organisms with similar characters due to common descent. The clade could be extended to include everything except the Lemurs, Lorises, and the lowest node. If this were the case, the line leading to lemurs would be considered the outgroup, while the rest of the primates would be considered the ingroup. These terms are simply used to describe different groups when discussing them in scientific writing.

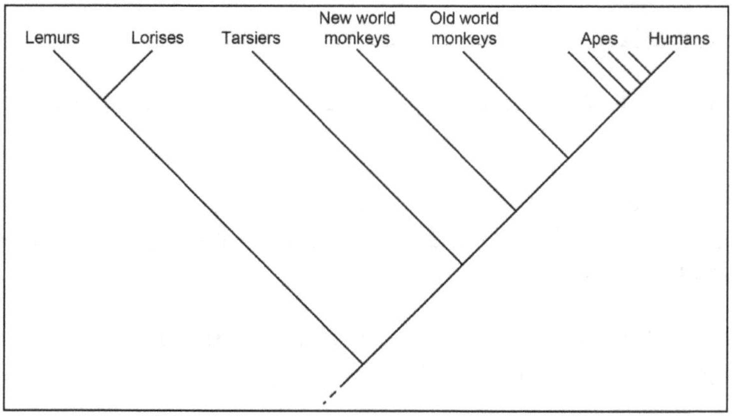

Cladogram of primates.

Interpreting Cladograms

In the following cladograms, it appears as if two different phylogenies are being presented. In the cladogram on the right, it appears that A is more closely related to C than in the cladogram on the left. This is simply a trick of presentation, but represents no meaning in terms of relatedness. These two cladograms do, in fact, represent a single phylogeny.

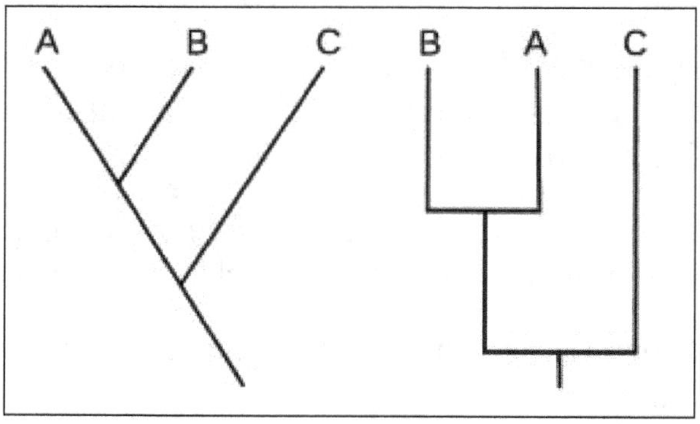

When creating or reading cladogram, it is important to remember that the only important features of the cladogram are the lines and the nodes. In these two cladograms, the lengths of the lines are roughly the same and more importantly, the nodes are in the same places. In both diagrams, A and B share a node that is further away from the origin of the line in the diagram. This tells us that A and B are more closely related than C is to either group. The order of A and B, as well as the orientation of the lines, does not matter. A cladogram may be drawn left to right, right to left, top to bottom, or bottom to top. Some large cladograms are even fashioned in a circle to include all the groups they represent. In some cladograms, evolutionary time in millions of years is represented to give an approximation from the lengths of the lines.

Phylogram

The phylogram R package is a tool for for developing phylogenetic trees as deeply-nested lists known as "dendrogram" objects. It provides functions for conversion between dendrograms and "phylo" class objects, as well as several tools for command-line tree manipulation and import/export via Newick parenthetic text. This improves accessibility to the comprehensive range of object-specific analytical and tree-visualization functions found across a wide array of bioinformatic R packages.

The R environment continues to gain popularity as a platform for bioinformatic analysis, due to the reproducible code-based workflow and the many powerful analytical tools available in a suite of open-source packages such as ape, phangorn and Phytools. These packages typically employ a tree structure known as the "phylo" object, whose primary element is an integer matrix with one row for each edge in the graph, and two columns giving the indices of the connecting nodes. This is a versatile and memory-efficient representation suitable for most applications encountered by evolutionary biologists, and hence a comprehensive array of tools has been developed for parsing, manipulating, analyzing and visualizing trees in matrix format.

An alternative tree structure is the "dendrogram" object, generated using the `as.dendrogram` function in the stats package. Rather than a matrix of edges, a dendrogram is a hierarchical list. These 'lists of lists' can be deeply nested, with the limit depending on the C stack size (settable via options("expressions")). While less memory-efficient than matrix-based tree objects, a useful feature of the nested-list representation is its modularity, whereby the sub tree of a tree is itself a tree - a dendrogram within a dendrogram. This means that dendrogram objects are subsettable in the same way that standard lists are, which in addition to the inbuilt editing functions such as `cut` and `merge`, facilitates intuitive command-line tree manipulation. An especially powerful feature of this object type is that tree editing operations can be carried out recursively using fast inbuilt functions in the "apply" family such as dendrapply and lapply.

Each node of a dendrogram object has the following mandatory attributes:

- "Height" the position of the node along the vertical axis (assuming the graph is orientated vertically).
- "Midpoint" the horizontal distance of the node from the left-most member of the sub tree (where the horizontal distance between adjacent leaves is 1 unit).
- "Members" the number of terminal leaf nodes belonging to the node.
- "Class" all nodes have the class attribute "dendrogram".

Rather than lists, terminal leaf nodes are length-1 integer vectors whose values correspond to the indices of the members in the set. The "members" attributes of leaf nodes is always 1, the "midpoint" attribute is 0, and they have two additional attributes:

- "Leaf" TRUE for terminal nodes (NULL otherwise).
- "Label" an optional character string giving the name of the taxon or group.

Aside from those listed above, users may attach other objects as attributes to the dendrogram nodes. For example, "label" attributes can be attached to inner nodes, and users can specify plotting parameters for each node by setting the attributes "nodePar" and "edgePar". Any number of non-standard objects can also be attached as node attributes, which can be useful for storing additional metadata such as bootstrap values and taxonomic information.

The flexibility, modularity and intuitive structure of dendrogram objects are appealing to many users, particularly where highly dynamic tree structures are required for applications such as machine learning clustering and classification. There is also a large and growing number of resources for manipulating and plotting dendrograms in contributed packages such as dendextend, and hence functions enabling bi-directional conversion between "dendrogram" and "phylo" class objects would expand the range of tools available for both object types. While conversion is currently possible using the "hclust" object as an intermediary, this object type does not support non-zero leaf node heights and hence is limited to ultrametric trees only.

Phylogram Package

Here, we introduce phylogram, an R package for structuring evolutionary trees as deeply-nested lists and transforming trees between list- and matrix-type objects. The package also contains functions for importing and exporting dendrogram objects to and from parenthetic text, as well as several functions for manipulating trees in nested-list format. These functions are detailed below with examples of their utility.

Importing and Exporting Trees

The Newick parenthetic text format is a universal phylogenetic tree representation that is compatible with most tree-editing software. The phylogram function read.dendrogram wraps the the Newick parser read.tree from the ape package, and converts the intermediate "phylo" object to a dendrogram. This function supports weighted edges, labels with special meta-characters (enclosed in single quotation marks), comments (enclosed in square brackets; ignored by the parser), multifuricating nodes, and both rooted and unrooted trees. Inner-node labels are also supported, and are attached as "label" attributes to non-leaf nodes. Objects of class "dendrogram" can also be exported as Newick-style parenthetic text using the function write.dendrogram.

Import and Export a Tree from a Newick String

Consider the simple example of a tree with three members named “A”, “B” and “C”, where “B” and “C” are more closely related to each other than either is to “A”. An unweighted Newick string for this tree would be (A,(B,C)); This can be imported as a dendrogram object using the read.dendrogram function as follows:

```
library(phylogram)

x <- read.dendrogram(text = "(A,(B,C));")

plot(x, yaxt = "n")
```

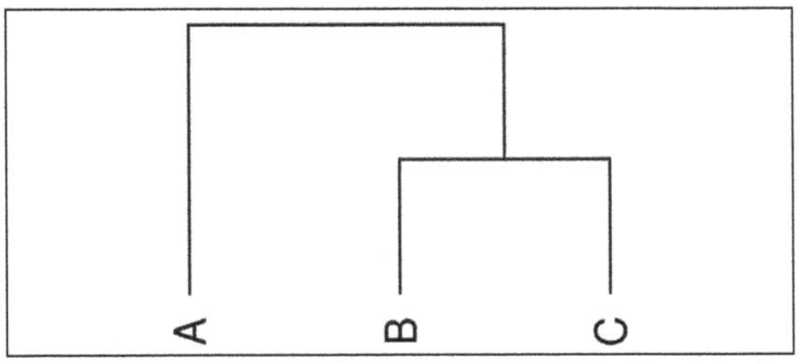

A simple dendrogram with three terminal leaf nodes.

The following command writes the object back to the console in Newick format without edge weights:

```
write.dendrogram(x, edges = FALSE)

#> "(A,(B,C));"
```

The syntax is similar when reading and writing text files, except that the text argument is replaced by file, and a valid file path is passed to the function.

Converting Tree Objects

Dendrograms can be converted to “phylo” objects and vice versa using the as.phylo.dendrogram and as.dendrogram.phylo methods. Unlike functions that employ an “hclust” object as an intermediary (e.g. as.dendrogram(as.hclust(phy))), these methods retain all weighted edges and do not require trees to be ultrametric. This facilitates access to the comprehensive range of functions that are specific for either dendrograms or “phylo” objects in contributed packages such as dendextend and ape, respectively. Note that other packages may employ the same function names, and hence the method dispatched may depend on the order in the which packages are loaded. For this reason it may be safer to use the full function call (e.g. phylogram::as.phylo.dendrogram(x) and phylogram::as.dendrogram.phylo(x)) when using these methods.

Convert a “Phylo” Object to a Dendrogram

A common application requiring conversion between “phylo” and “dendrogram” objects involves plotting tanglegrams to visualize incongruence between two phylogenetic trees. The dendextend package features the function tanglegram for versatile plotting of two distinct trees and indicating the discordant nodes using a series of non-parallel edges. However this function does not currently support non-ultrametric "phylo" objects. In this example, two weighted neighbor-joining trees are generated from the left and right sections of the woodmouse alignment from the ape package, and converted to dendrograms for visual comparison using dendextend::tanglegram.

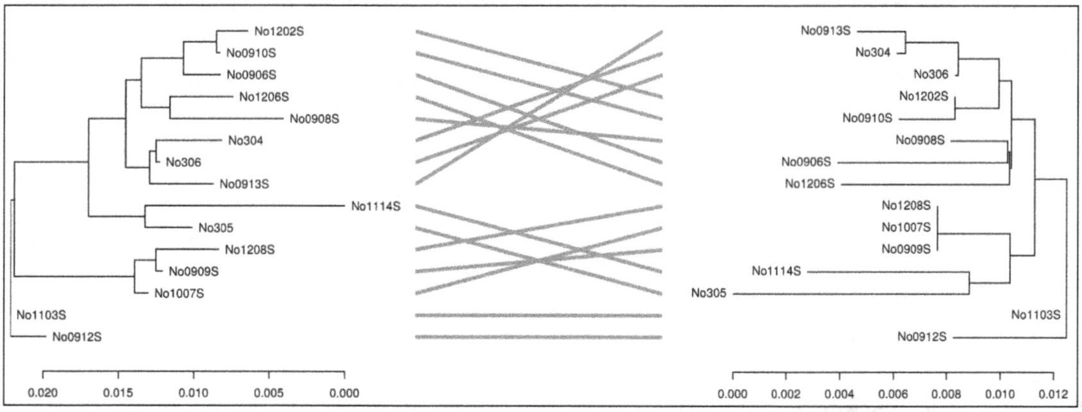

Tanglegram showing incongruence between the left- and right-hand sections of the woodmouse alignment.

```
library(ape)

data(woodmouse)

## generate distance matrices for each section of the alignment

dist1 <- dist.dna(woodmouse[, 1:482])

dist2 <- dist.dna(woodmouse[, 483:965])

## build neighbor-joining trees

phy1 <- nj(dist1)

phy2 <- nj(dist2)

## root with No0912S as outgroup

phy1 <- root(phy1, "No0912S")

phy2 <- root(phy2, "No0912S")

## convert phylo objects to dendrograms

dnd1 <- as.dendrogram(phy1)
```

```
dnd2 <- as.dendrogram(phy2)

## rearrange in ladderized fashion

dnd1 <- ladder(dnd1)

dnd2 <- ladder(dnd2)

## plot the tanglegram

dndlist <- dendextend::dendlist(dnd1, dnd2)

dendextend::tanglegram(dndlist, fast = TRUE, margin_inner = 5)
```

Tree Editing/Manipulation

The phylogram package features several additional functions to facilitate some of the more common manipulation operations. Leaf nodes and internal branching nodes can be removed using the function `prune`, which identifies and recursively deletes nodes based on pattern matching of "label" attributes. This is slower than the ape function `drop.tip`, but offers the benefits of versatile string matching using regular expressions, and the ability to remove inner nodes (and by extension all of their subnodes) that feature matching "label" attributes. To aid visualization, the function ladder rearranges the tree, sorting nodes by the number of members (analogous to the `ladderize` function in the ape package). Another function aiding in tree visualization is `as.cladogram`, which resets the "height" attributes of all terminal leaf nodes to zero and progressively resets the heights of the inner nodes by single incremental units in a bottom-up fashion. The function `reposition` scales the heights of all nodes in a tree by a given constant (passed via the argument `shift`), and features the option to reset all node heights so that height of the farthest terminal leaf node from the root is zero (by specifying `shift = "reset"`). The function remidpoint recursively corrects all "midpoint", "members" and "leaf" attributes following manual editing of a tree or while converting a nested list to a "dendrogram" object.

Building and Manipulating Dendrograms

The simple three-leaf dendrogram in figure can be created manually as follows:

```
x <- list(1, list(2, 3))

## attach "leaf" and "label" attributes to leaf nodes

attr(x[], "leaf") <- TRUE

attr(x[][], "leaf") <- attr(x[][], "leaf") <- TRUE

attr(x[], "label") <- "A"

attr(x[][], "label") <- "B"
```

```
attr(x[][], "label") <- "C"
## set "height" attributes for all nodes
attr(x, "height") <- 2
attr(x[], "height") <- 0
attr(x[], "height") <- 1
attr(x[][], "height") <- attr(x[][], "height") <- 0
## set "midpoints" attributes for all nodes
attr(x, "midpoint") <- 0.75
attr(x[], "midpoint") <- 0
attr(x[], "midpoint") <- 0.5
attr(x[][], "midpoint") <- attr(x[][], "midpoint") <- 0
## set "members" attributes for all nodes
attr(x, "members") <- 3
attr(x[], "members") <- 1
attr(x[], "members") <- 2
attr(x[][], "members") <- attr(x[][], "members") <- 1
## set class as "dendrogram"
## Note that setting the class for the root node
## automatically sets the class of all nested subnodes
class(x) <- "dendrogram"
x
#> 'dendrogram' with 2 branches and 3 members total, at height 2
```

As demonstrated above, manually setting attributes on dendrogram objects can be rather tedious, motivating the development of functions to automate the generation and manipulation of these tree structures.

This simple tree can be recreated more succinctly using the phylogram package functions as follows:

```
x <- list(1, list(2, 3))
## recursively set class, midpoint, members and leaf attributes
x <- remidpoint(x)
## set incremental height attributes
x <- as.cladogram(x)
```

```
## set label attributes using dendrapply
set_label <- function(node){
 if(is.leaf(node)) attr(node, "label") <- LETTERS[node]
 return(node)
}
x <- dendrapply(x, set_label)
x
#> 'dendrogram' with 2 branches and 3 members total, at height 2
```

Similarly, dendrogram objects can be subset using either the prune function or standard list-subsetting syntax, again with the help of utility functions to recursively reset node attributes. The following code demonstrates one option for rearranging the tree with species A and B as sister taxa and C as the ancestor:

```
## isolate root node (species C)
ancestor <- prune(x, pattern = "C", keep = TRUE)
## alternative option using subset operator
ancestor <- x[][]
## create subtree without species C
subtree <- prune(x, pattern = "C")
## graft subtree onto root
x <- list(ancestor, subtree)
## set attributes as above
x <- as.cladogram(remidpoint(x))
## plot dendrogram
plot(x, yaxt = "n")
```

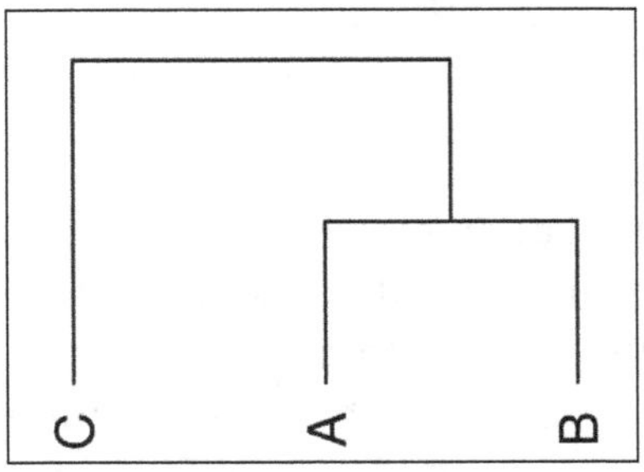

Rearranged dendrogram with species C ancestral to A and B.

Tree Visualization

Publication-quality trees can be generated from dendrogram objects using the stats plotting function `plot.dendrogram`, and the extensive plotting functions available in dendrogram-enhancing packages such as circlize and dendextend. The latter also offers the facility to convert dendrograms to "ggdend" objects, for which many powerful 'grammar of graphics' plotting functions are available in the ggplot2 and ggdendro packages. Moreover, there are several advanced plotting options for "phylo" objects in the ape package, as well as the Bioconductor package ggtree. Given the extensive tree visualization options already available, we do not include any additional plotting functions in the phylogram package.

PHYLOGENETIC NETWORK

A phylogenetic network is any graph used to visualize evolutionary relationships (either abstractly or explicitly) between nucleotide sequences, genes, chromosomes, genomes, or species. They are employed when reticulation events such as hybridization, horizontal gene transfer, recombination, or gene duplication and loss are believed to be involved. They differ from phylogenetic trees by the explicit modeling of richly linked networks, by means of the addition of hybrid nodes (nodes with two parents) instead of only tree nodes (a hierarchy of nodes, each with only one parent). Phylogenetic trees are a subset of phylogenetic networks. Phylogenetic networks can be inferred and visualised with software such as SplitsTree, the R-package, phangorn, and, more recently, Dendroscope. A standard format for representing phylogenetic networks is a variant of Newick format which is extended to support networks as well as trees.

Many kinds and subclasses of phylogenetic networks have been defined based on the biological phenomenon they represent or which data they are built from (hybridization networks, usually built from rooted trees, recombination networks from binary sequences, median networks from a set of splits, optimal realizations and reticulograms from a distance matrix), or restrictions to get computationally tractable problems (galled trees, and their generalizations level-k phylogenetic networks, tree-child or tree-sibling phylogenetic networks).

Microevolution

Phylogenetic trees also have trouble depicting microevolutionary events, for example the geographical distribution of muskrat or fish populations of a given species among river networks, because there is no species boundary to prevent gene flow between populations. Therefore, a more general phylogenetic network better depicts these situations.

Rooted vs. Unrooted

Unrooted Phylogenetic Network

Let X be a set of taxa. An unrooted phylogenetic network N on X is any undirected graph whose leaves are bijectively labeled by the taxa in X. A number of different types of unrooted phylogenetic networks are in use like split networks and quasi-median networks. In most cases, such networks only depict relations between taxa, without giving information about the evolutionary history. Although some methods produce unrooted networks that can be interpreted as undirected versions of rooted networks, which do represent a phylogeny.

Rooted Phylogenetic Network

Let X be a set of taxa. A rooted phylogenetic network N on X is a rooted directed acyclicgraph where the set of leaves is bijectively labeled by the taxa in X. Rooted phylogenetic networks, like rooted phylogenetic trees, give explicit representations of evolutionary history. This means that they visualize the order in which the species diverged (speciated), converged (hybridized), and transferred genetic material (horizontal gene transfer).

Classes of Networks

For computational purposes, studies often restrict their attention to classes of networks: subsets of all networks with certain properties. Although computational simplicity is the main goal, most of these classes have a biological justification as well. Some prominent classes currently used in the mathematical phylogenetics literature are tree-child networks, tree-based networks, and level-k networks.

BAYESIAN NETWORK

Gene interaction (GI) networks provide insight for understanding the biological mechanisms that explain various phenotypes in health and disease. The inference of GI networks from high-throughput biological data is an important and challenging task in systems biology. Throughout the literature, the term GI has been used in a broad sense implying direct and indirect interactions between genes and gene products. Several machine learning and statistical methods have been proposed for the problem and Bayesian network (BN) models have gained popularity for the task of inferring gene networks. Because of the complexity of GI networks and the sparse, noisy nature of experimental data, machine learning and statistical methods may lead to poor reconstruction accuracy for the underlying network. One way to overcome this problem would be to incorporate prior biological knowledge when making network inferences using experimental data. Due to technological advances in sequencing, microarray, proteomics and related fields, biological and clinical data are being produced at an ever increasing rate.

BNs have a number of features that make them viable candidates for combining prior knowledge and data as BNs can deal with uncertainty, avoid over fitting a model to training data, and learn from incomplete datasets. BNs handle stochastic events in a probabilistic framework accounting for noise, which results in emphasizing only strong relations in the observed data. Furthermore, BNs are able to focus on local interactions where each node is directly affected by a relatively small number of nodes and interactions defined by a BN can be related to causal inference. These properties are similarly observed in biological networks justifying the use of BNs in exploring pathways in the setting of identifying GI networks using experimental data. Learning algorithms for both the structure and parameters of BNs have been developed. Most of the research on BNs has focused on directed acyclic graphs (DAGs) and static systems with discrete variables and linear Gaussian models. Friedman et al. used BNs to generate a causal model of the yeast cell-cycle data using either a model with discretized expression levels (e.g. Boolean, or underexpressed/normal/overexpressed), or a linear Gaussian model. The latter treats the expression level of a gene as being normally distributed around a mean which is a linear sum of inputs. Therefore, rather than true causal relationships, the results may represent co-regulation of genes. Accordingly, a method to sample network structures from the posterior distribution with Markov Chain Monte Carlo (MCMC) has been introduced.

Many BN structure learning algorithms are based on heuristic search techniques with the likelihood approximation because of the infeasible computational complexity. These approaches may lead to a false model, as neither the search technique nor the objective functions guarantee the optimal solution. Informative priors generated from existing biological information can improve structure learning to get better models to describe the underlying GIs. In several studies the use of prior biological knowledge in conjunction with gene-expression data has been shown to improve the fidelity of network reconstruction. Hartemink et al. incorporated genomic location data to guide the BN model inference. Tamada et al. proposed a method, which iteratively detects consensus motifs based on the structure of the estimated network model, then evaluates the network using the result of the motif detection, until the inferred network becomes stable. Imoto et al. proposed a framework utilizing Gibbs distribution where an energy function was used to evaluate the probability of an edge in the inferred networks. Werhli and Husmeier extended this approach to integrate multiple sources of prior knowledge into dynamic Bayesian network (DBN) learning via MCMC sampling. Mukherjee and Speed proposed a scheme to incorporate known network features including edges, classes of edges, degree distributions, and sparsity into gene network reconstruction within a Bayesian learning framework utilizing MCMC sampling.

These studies were limited in the use of external biological knowledge by incorporating only certain features, such as network topology or binding sites in promoter regions. Furthermore, in the aforementioned approaches manual curation and incorporation of the external knowledge are employed. In this article, we present a framework to incorporate multiple sources of prior knowledge, regardless of its type, into BN learning. The meaning of prior knowledge in our context is the enumeration of pair-wise interactions of genes from biological information sources and the use of this information in BN modeling. The proposed method is fully automatic and does not use likelihood approximations to find the optimal network that explains observed experimental data. We propose a novel framework that uses BN infrastructure itself to incorporate external biological knowledge when learning networks. This infrastructure yields GI information for pairs of genes, which can be used as informative priors to calculate the probability of a candidate graph, G. This information is then incorporated in the network-learning process that tries to identify the most probable graph given data.

Methods

The schematic depiction of the overall proposed method is presented in figure. Pairwise interaction information is gathered from biological databases and a BN model for prior knowledge, Bayesian Network Prior (BNP) is developed. In BNP, one node is depicted as GI and the topology represents the dependence structure for different evidence types within each other and with the GI node. For a set of genes, the model is instantiated with the given evidence and experimental data for each pair of genes. The GI node is used to infer whether the gene pair is related or not, represented by a prediction value between 0 and 1. A prior knowledge matrix, B, is populated with these prediction values for all gene pairs. Using a proposed novel energy formula and informative prior formula, this prior knowledge is utilized to calculate the probability of a candidate DAG, G, in the structure learning process. This parameter is used to optimize P(G|D) instead of the likelihood, P(D|G), used by existing structure learning algorithms.

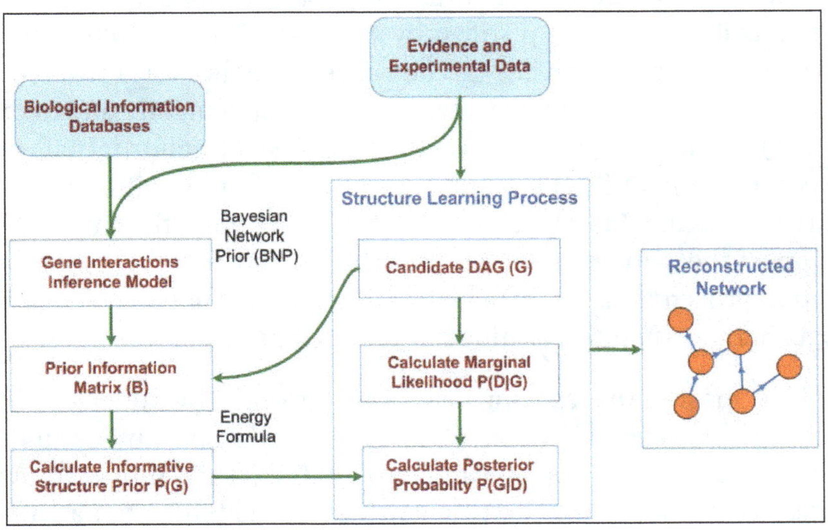

Overall workflow of the proposed method. BNP is constructed using GI information from external biological databases and when instantiated with an evidence vector for a pair of genes, the GI probability is inferred. For a list of genes, the pair-wise interaction information is stored in the prior matrix B, which is used to calculate the probability of a candidate graph G in the structure learning process.

Informative Structure Priors

A BN is a compact graphical representation of the joint probability distribution over a set of random variables and consists of a DAG = (V, E), with a node set V corresponding to the random variables $X_1, \ldots, X_n$ and an edge set E on these nodes and a set of conditional probability distributions Θ for each node in the DAG. The DAG encodes the assertions of conditional independence. If the random variables are discrete, conditional probability distributions Θ can be represented as a set of conditional probability tables (CPT). CPTs list the probabilities for each value that a child node can assume given a combination of values of its parents.

In GI network-modeling studies using BNs, X_i represents a gene and edges represent relationship between genes. The task of network inference (i.e. structure learning) is to make inferences regarding the graph G that best explains the data. This can be achieved by finding the DAG G that maximizes:

$$P(G|D) = \frac{P(G|D)P(G)}{P(H)}.$$

where P(D|G) is the likelihood, P(D) is the probability of the data, P(G) is the structure prior (or network prior) probability of the graph G and P(G|D) is the posterior probability of G. In commonly used heuristic structure learning algorithms, P(D|G) is optimized instead of the true model P(G|D). The likelihood criterion does not guarantee to find the optimum solution even if a heuristic approach is not employed. Nevertheless, optimizing the likelihood can be justified by assuming P(D) and P(G) to be equal for all G. The former assumption can be regarded as reasonable as D is observed. However, the latter assumption is generally not correct and is made mainly due to difficulties in calculating P(G) and lack of prior knowledge on G. Use of uniform (flat) priors for Gs ignores the contribution of P(G) and this may cause failure in differentiating between DAGs that are in the same Markov equivalence set. Therefore, the true DAG among the ones that support the same conditional probability distribution cannot be identified. The proposed approach aims to calculate P(G) using external knowledge and provide improvements in the structure-learning phase for GI networks.

For discrete BNs, most of the learning tasks are performed by calculating P(D|G) with the Bayesian Dirichlet equivalent (BDe) scoring function and by assuming uniform (flat) prior structure for all possible candidate DAGs. In the proposed approach, we employ a greedy search algorithm that aims to maximize P(G|D). For a given candidate

DAG, G, we calculate P(G) by first obtaining the prior information matrix, B. Unlike existing methods, the proposed approach does not use categorized prior knowledge but assigns probabilities to each candidate edge. The matrix B is obtained by instantiating BNP with the evidence vector for each pair of genes in the input gene set. These evidence vectors can originate from any performed experimental data at hand, or external knowledge, or both.

Let B be the prior information matrix, where B(i, j) = $P(X_{ij})$, the degree of prior belief that gene i and j interact based on external knowledge. Let A_G denotes the adjacency matrix of the candidate graph G. We define the matrix U such that U(i, j) = 1 − [B(i, j) A_G(i, j)], the element by element multiplication of B and A_G. Note that if there exists no edge from i to j in G, U(i, j) = 1; and if there is an edge from i to j in G, U(i, j) is inversely proportional to our prior belief on the existence of the edge. The total energy of G is defined as:

$$E(G)=\sum_{i,j}\frac{U(i,j)}{N^2},$$

where N is the number of nodes in G. This way, we do not assign categorical values for U(i, j) and exploit fully the information about prior existence of an edge. Informative structure prior is formulated as:

$$P(G\,|\,\beta)=Ce^{-\beta E(G)},$$

where C is a scaling constant. The choice of C does not affect the relative comparison during scoring of graphs in structure learning. The hyperparameter β can be marginalized using:

$$P(G)=c.\frac{1}{\beta_H-\beta_L}\int_{\beta_L}^{\beta_H}e^{-\beta E(G)}d\beta.$$

For ease of simulation, the integral is calculated for a range of E(G) and stored in a lookup table. In the numerical calculation of this integral, Δβ = βH − βL is the parameter of interest, the integration approach automatically incorporates the uncertainty on the parameter by averaging the likelihood values of the parameter. Point estimates acquired by maximizing the parameters may change arbitrarily with arbitrary re-parameterizations. Point estimates maximize the probability density without taking into account the complementary volume information, which may yield in suboptimal results. When one has a choice of which variables to integrate over and which to maximize over, it is suggested that one would integrate over as many variables as possible in order to capture the relevant volume information of high-dimensional probability distributions.

Bayesian Network Prior

The goal in building BNP is to construct a framework such that the distilled external biological knowledge is used in an intelligent way to make an assessment about the interaction of a pair of genes. Previously, Troyanskaya et al. proposed a Bayesian Framework for combining various data sources for gene function prediction. In this method a Naive Bayesian model was constructed. The parameters (CPTs) of the model were determined by experts. Then, a separate network was instantiated for each gene pair by initializing the bottom-level nodes with evidence and the probability of the functional relationship between the two genes was updated. The model was designed for functional prediction, not for GI network learning. Here, we describe a novel-prior knowledge inference model that automatically learns parameters of the nodes used in BNP that predicts if two genes interact using external biological knowledge. The model organism chosen for BNP was human and the external data came from pathway, microarray, gene and protein interaction databases. The assembled information source is made up of 'evidence types', each making a 'Yes' or 'No' call about the interaction of two genes and BNP is the BN that represents the relation between these evidence types and GI.

Microarray co-expression was calculated using two datasets. The first dataset aims to provide a gene atlas for the human genes and examines 79 normal human tissues with 158 samples. The second database came from the 'Reference Database for Gene Expression Analysis' (RefExA) that represents 70 normal human tissue samples. Affymetrix Expression Console v1.1 was used to normalize the samples using the MAS 5.0 method. Probe sets with absence calls in all of the samples were omitted from further analysis. Centered Pearson correlations were calculated and 71 617 pairs of probe sets with a correlation value greater than 0.98 were passed on to be used to construct BNP. KEGG, NCI/NATURE and Reactome databases were utilized to gather pathway based evidence data. 3258 pair-wise gene relations that existed in at least two of the three pathway databases were used for further analysis. A dataset was obtained from BioGrid with evidence of interactions that are observed in experiments with 17 different assay types such as affinity capture and two-hybrid. BioGrid analysis revealed a total of 35 600 non-redundant pair-wise interactions. After the microarray probe set level data was regressed to the gene level and all three sources were merged, 60 950 pair-wise GIs based on 19 evidence types were obtained. A GI node is appended to this evidence matrix (where rows represent gene pairs and columns represent evidence types) with a 'true' value if there were at least two evidence types implying interaction. BNP was built by learning both structure and parameters using Greedy Hill Climbing.

Applications of Bayesian Network

Gene Regulatory Network

GRN is Gene Regulatory Network or Genetic Regulatory Network. It comprises of several DNA segments in a cell. It interacts with other substances in the cell and also with each other indirectly. Indirectly means through their protein and RNA expression products. Thus, it governs the expression levels of mRNA and proteins. GRNs reproduce the behaviour of the system using Mathematical models. In some cases, corresponding with experimental observations, it generates predictions.

Medicine

It is the science or practice of diagnosis. For the treatment and prevention of any disease, we use medicines. We are using medicines since ancient times. Over the years, medicines and drugs have evolved to cater to a variety of health care practices. In order to provide better healthcare, machines and other computer devices assist us in the diagnosis of the disease.

Biomonitoring

We use biomonitoring to quantify the concentration of chemicals. It measures the concentration in blood and tissue of humans, etc. Hence, it is the measurement of the body burden in analytical chemistry. Biomonitoring involves the use of indicators. These measurements are often done in blood and urine. To determine the levels of many ECCs in humans, DTSC scientist is conducting biometric studies.

Document Classification

It is a problem in library science, computer science, and information science. The main task is to assign a document to multiple classes. We can also do it manually or algorithmically. Manual classification is intellectual classification and it takes time. We use the algorithmic classification of documents in information science and computer science.

Information Retrieval

It is the activity of obtaining information resources. Information retrieval concerns retrieving the information from databases. It is a continuous process. During the process, we can consider, reconsider and refine our research problem. Metadata or full-text indexing is the basis of searching. To reduce "information overload", we use automated information retrieval systems.

Semantic Search

By understanding searcher intent and the contextual meaning of terms, it improves search accuracy. It enhances the accuracy in the searchable dataspace, whether on the web or within a closed system, to generate more relevant results.

Image Processing

It is the processing of images by using mathematical operations. We can also use image processing to convert images into digital format. After converting the images, we can also apply some operations on it to enhance the image. Image processing is any form of signal processing.

In this, the input can be formed as an image, such as a photograph or video frame. The output of image processing may be either a set of characteristics or parameters related to the image or an image. Hence, in image processing techniques, we generally treat the image as a two-dimensional signal. After that, we apply standard signal processing on it.

Spam Filter

The spam filter is a program. We use a spam filter to detect unsolicited and unwanted email. Bayesian spam filter calculates whether the message is spam or not. The Bayesian spam filter is more robust than other spam filters. We use filtering to learn from spam and ham messages.

Turbo Code

The turbo codes are a class of high-performance forward error correction codes. Thus, turbo code uses the Bayesian Network. Turbo codes are the state of the art of codecs. 3G and 4G mobile telephony standards use these codes. Hence the Bayesian Network represents turbo coding and decoding process.

System Biology

We can also use BN to infer different types of biological network from Bayesian structure learning. In this, the main output is the qualitative structure of the learned network.

APPLICATIONS OF PHYLOGENETICS

- Classification: Phylogenetics based on sequence data provides us with more accurate descriptions of patterns of relatedness than was available before the advent of molecular sequencing. Phylogenetics now informs the Linnaean classification of new species.

- Forensics: Phylogenetics is used to assess DNA evidence presented in court cases to inform situations, e.g. where someone has committed a crime, when food is contaminated, or where the father of a child is unknown.

- Identifying the origin of pathogens: Molecular sequencing technologies and phylogenetic approaches can be used to learn more about a new pathogen outbreak. This includes finding out about which species the pathogen is related to and subsequently the likely source of transmission. This can lead to new recommendations for public health policy.

- Conservation: Phylogenetics can help to inform conservation policy when conservation biologists have to make tough decisions about which species they try to prevent from becoming extinct.
- Bioinformatics and computing: Many of the algorithms developed for phylogenetics have been used to develop software in other fields.

References

- Phylogenetics, science: britannica.com, Retrieved 17 April, 2020
- Baum DA, Smith SD (2013). Tree Thinking: An Introduction to Phylogenetic Biology. Roberts. p. 442. ISBN 978-1-936221-16-5
- What-phylogenetics, introduction-phylogenetics, training: ebi.ac.uk, Retrieved 26 July, 2020
- Phylogram-vignette, phylogram, vignettes, packages: cran.r-project.org, Retrieved 29 August, 2020
- Wiens JJ (February 2006). "Missing data and the design of phylogenetic analyses". Journal of Biomedical Informatics. 39 (1): 34–42. doi:10.1016/j.jbi.2005.04.001. PMID 15922672
- Bayesian-network-applications: data-flair.training, Retrieved 20 March, 2020

Biological Database

CHAPTER 4

Biological database refers to the stores of biological information. It includes genomics, proteomics, metabolomics, microarray gene expression, gene function and mutation. This chapter has been carefully written to provide an easy understanding of the related concepts of biological database.

Bioinformatics is the application of Information technology to store, organize and analyze the vast amount of biological data which is available in the form of sequences and structures of proteins (the building blocks of organisms) and nucleic acids (the information carrier). The biological information of nucleic acids is available as sequences while the data of proteins is available as sequences and structures. Sequences are represented in single dimension whereas the structure contains the three dimensional data of sequences.

A biological database is a collection of data that is organized so that its contents can easily be accessed, managed, and updated. The activity of preparing a database can be divided in to:

- Collection of data in a form which can be easily accessed.
- Making it available to a multi-user system (always available for the user).

```
                                                    N
EXPERIMENTS                                         E   U
     |                                           |- T-->S            COPY
ORGANIZATION                         |-----------|- W-->E-->ONLINE -----> PERSONAL
  OF DATA                            HOST/SERVER |- O-->R   ACCESS        DATABASE
     |                                   |          R   S
     |------------->  DATABASES                     K
                          |
                         EDS
                (Electronic Data Storage)
```

The network for production, construction and accession of a database.

Biological databases can be broadly classified in to sequence and structure databases. Sequence databases are applicable to both nucleic acid sequences and protein sequences, whereas structure database is applicable to only Proteins. The first database was created within a short period after the Insulin protein sequence was made available in 1956. Incidentally, Insulin is the first protein to be sequenced. The sequence of Insulin consisted of just 51 residues (analogous to alphabets in a sentence) which characterize the sequence. Around mid-nineteen sixties, the first nucleic acid sequence of Yeast tRNA with 77 bases (individual units of nucleic acids) was found out. During this period, three dimensional structures of proteins were studied and the well-known Protein Data Bank was developed as the first protein structure database with only 10 entries in 1972. This has now grown in to a large database with over 10,000 entries. While the initial databases of protein sequences were maintained at the individual laboratories, the development of a consolidated formal database known as SWISS-PROT protein sequence database was initiated in 1986 which now has about 70,000 protein sequences from more than 5000 model organisms, a small fraction of all known organisms. These huge varieties of divergent data resources are now available for study and research by both academic institutions and industries. These are made available as public domain information in the larger interest of research community through Internet and CDROMs. These databases are constantly updated with additional entries.

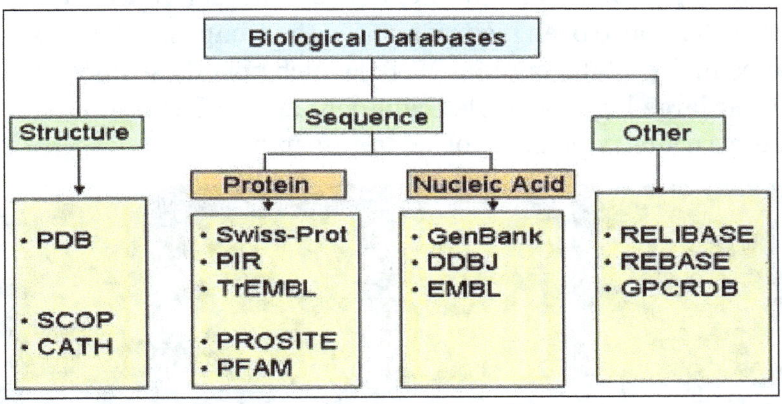

Biological Databases.

Databases in general can be classified in to primary, secondary and composite databases. A primary database contains information of the sequence or structure alone. Examples of these include Swiss-Prot and PIR for protein sequences, GenBank & DDBJ for Genome sequences and the Protein Databank for protein structures.

Primary databases	Secondary databases
• Original submission by the experimentalists who have originally searched	• Built up from primary data which is retrieved by primary database
• Content Is controlled by the submitters	• Content controlled by third party NCBI
• Examples include GENEBANK, SNP and GEO	• Examples include RefSeq, RefSNP, NCBI Structure, Protein. Etc.

A secondary database contains derived information from the primary database. A secondary sequence database contains information like the conserved sequence, signature sequence and active site residues of the protein families arrived by multiple sequence alignment of a set of related proteins. A secondary structure database contains entries of the PDB in an organized way. These contain entries that are classified according to their structure like all alpha proteins, all beta proteins, etc. These also contain information on conserved secondary structure motifs of a particular protein. Some of the secondary database created and hosted by various researchers at their individual laboratories includes SCOP, developed at Cambridge University; CATH developed at University College of London, PROSITE of Swiss Institute of Bioinformatics, eMOTIF at Stanford.

Composite database amalgamates a variety of different primary database sources, which obviates the need to search multiple resources. Different composite database use different primary database and different criteria in their search algorithm. Various options for search have also been incorporated in the composite database. The National Center for Biotechnology Information (NCBI) which host these nucleotide and protein databases in their large high available redundant array of computer servers, provides free access to the various persons involved in research.

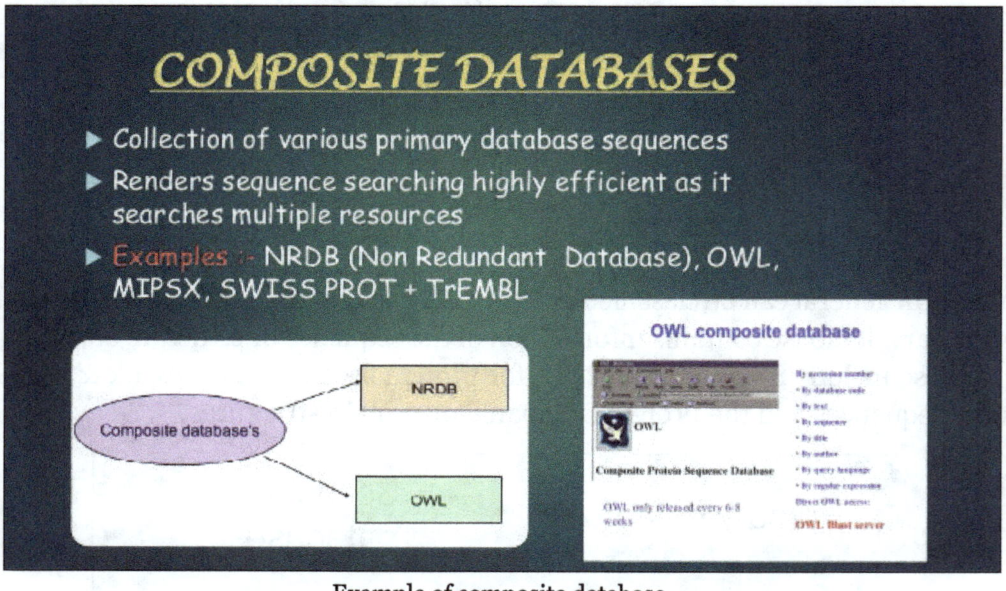

Example of composite database.

The growth of the primary databases gave rise to serious and valid questions on the format of the sequences, reliability and the comprehensiveness of the databases. To address the format issues, in-house software solutions have been developed to convert format of one database to another. A public domain software FORCON can also be used. The newer software tools which are used for analysis accept the data in multiple formats. The problem in the reliability of the data is the possibility of misannotations. The misannotations are some time introduced due to the process of automation of annotation process which is carried out extensively with the help of computers.

Misannotations, if introduced, multiply in subsequent additions and may accumulate to an unbelievable extent and create confusion. A possible solution to prevent this from happening is to flag the protein sequence which has been annotated by sequence comparison but whose function has not been validated by experimental methods.

Protein Sequence Analysis

Apart from maintaining the large database, mining useful information from these set of primary and secondary databases is very important. Lot of efficient algorithms have been developed for data mining and knowledge discovery. These are computation intensive and need fast and parallel computing facilities for handling multiple queries simultaneously. It is these search tools that integrate the user and the databases. One of the widely used search program is BLAST (Basic Local Alignment Search Tool).

BLAST is a set of similarity search programs designed to explore all of the available sequence databases regardless of whether the query is protein or DNA. The BLAST programs have been designed for speed, with a minimal sacrifice of sensitivity. The scores assigned in a BLAST search have a well-defined statistical interpretation, making real matches easier to distinguish from random background hits. BLAST uses a heuristic algorithm which seeks local as opposed to global alignments and is therefore able to detect relationships among sequences which share only isolated regions of similarity. This is a primary criterion in sequence analysis. Other tool available includes CLUSTALW for multiple sequence alignment.

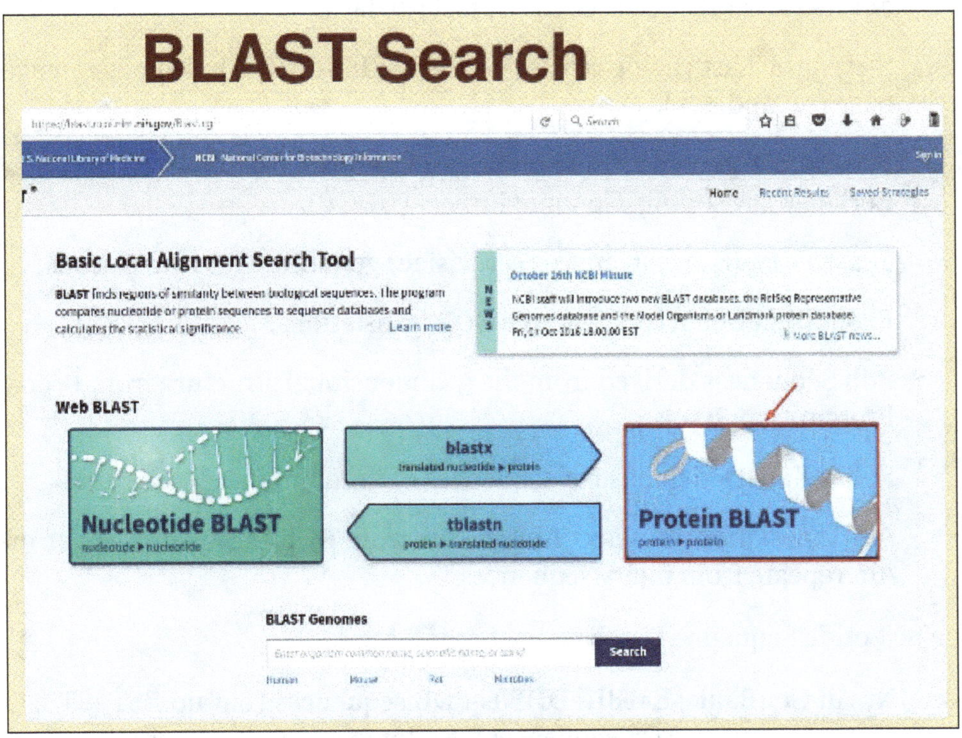

BLAST - Basic Local Alignment Search Tool.

Programs available for the BLAST search include the following:

- Blastp compares an amino acid query sequence against a protein sequence database.
- Blastn compares a nucleotide query sequence against a nucleotide sequence database.
- Blastx compares a nucleotide query sequence translated in all reading frames against a protein sequence database.
- Tblastn compares a protein query sequence against a nucleotide sequence database dynamically translated in all reading frames.
- Tblastx compares the six-frame translations of a nucleotide query sequence against the six-frame translations of a nucleotide sequence database. Note that tblastx program cannot be used with the nr database on the BLAST Web page.

Databases available for BLAST search include the following:

- Protein Sequence Databases:
    - Nr All non-redundant GenBank CDS translations+PDB+SwissProt+PIR+PRF.
    - Month All new or revised GenBank CDS translation+PDB+SwissProt+PIR+PRF released in the last 30 days.
    - Swissprot Last major release of the SWISS-PROT protein sequence database (no updates).
    - Drosophila genome Drosophila genome proteins provided by Celera and Berkeley Drosophila Genome Project (BDGP).
    - Yeast Yeast (Saccharomyces cerevisiae) genomic CDS translations.
    - Ecoli Escherichia coli genomic CDS translations.
    - Pdb Sequences derived from the 3-dimensional structure from Brookhaven Protein Data Bank.
    - Kabat Kabat's database of sequences of immunological interest.
    - Alu Translations of select Alu repeats from REPBASE, suitable for masking Alu repeats from query sequences.
- Nucleotide Sequence Databases:
    - Nr All GenBank+EMBL+DDBJ+PDB sequences (but no EST, STS, GSS, or phase 0, 1 or 2 HTGS sequences). No longer "non-redundant".

- Month All new or revised GenBank+EMBL+DDBJ+PDB sequences released in the last 30 days.
- Drosophila genome Drosophila genome provided by Celera and Berkeley Drosophila Genome Project.
- Dbest Database of GenBank+EMBL+DDBJ sequences from EST Divisions.
- Dbsts Database of GenBank+EMBL+DDBJ sequences from STS Divisions.
- Htgs Unfinished High Throughput Genomic Sequences: phases 0, 1 and 2.
- Gss Genome Survey Sequence, includes single-pass genomic data, exon-trapped sequences, and Alu PCR sequences.
- Yeast Yeast (Saccharomyces cerevisiae) genomic nucleotide sequences.
- E. coli Escherichia coli genomic nucleotide sequences.
- Pdb Sequences derived from the 3-dimensional structure from Brookhaven Protein Data Bank.
- Kabat Kabat's database of sequences of immunological interest.
- Vector Vector subset of GenBank(R).
- Mito Database of mitochondrial sequences.
- Alu Select Alu repeats from REPBASE, suitable for masking Alu repeats from query sequences.
- Epd Eukaryotic Promotor Database found on the web.

ENSEMBLE CLUSTERING FOR BIOLOGICAL DATASETS

Biological data requires both low and high level analysis to reveal significant information that will shed light into biological facts such as disease prediction, annotation of a gene function and guide new experiments. In that sense, researchers are seeking for the effect of a treatment or time course change befalling. For example, they may design a microarray experiment treating a biological organism with a chemical substance and observe gene expression values comparing with expression value before treatment. This treatment or change make researchers focus on groups of genes, other biological molecules that have significant relationships with each other under similar conditions. For instance, gene class labels are usually unknown, since there is a little information available about the data. Hence, data analysis using an unsupervised learning technique is required. Clustering is an unsupervised learning technique used in diverse domains including bioinformatics. Clustering assigns objects into the same cluster, based on a cluster definition. A cluster definition or criterion is the similarity between the objects. The idea is that one needs to find the most important cliques among many from the data. Therefore, clustering is widely used to obtain biologically meaningful partitions. However, there is no best clustering approach for the problem on hand and clustering algorithms are biased towards certain criteria. In other words, a particular clustering approach has its own objective and assumptions about the data.

Diversity of clustering algorithms can benefit from merging partitions generated individually. Ensemble clustering provides a framework to merge individual partitions from different clustering algorithms. Ensemble clustering may generate more accurate clusters than individual clustering approaches. Here, an ensemble clustering framework is implemented to aggregate results from K-means, hiearchical clustering and C-means algorithms. We also use different data sets. Two different biological datasets are used for each algorithm. A comparison of the results is presented. In order to evaluate the performance of the ensemble clustering approach, one internal and one external cluster validation indices are used. Silhouette is the internal validation index and C-rand is the external one.

Clustering Algorithms

Clustering biological data is very important for identification of co-expressed genes, which facilitates functional annotation and the elucidation of biological pathways. Accurate predictions can serve as a guide for targeting further experiments and generating additional hypotheses. Furthermore, accurate predictions can facilitate identification of disease markers and targets for drug design; clustering can also be used to determine whether certain patterns exist near viral integration sites.

Current algorithms used in gene clustering have some drawbacks. For example, K-means algorithm is sensitive to noise that is inherent in gene expression data. In addition, the solution (i.e. the final clustering) that the K-means algorithm finds may not be a global optimum since it relies on randomly chosen initial objects. K-means works upon randomly chosen centroid points that represent the clusters. The objects are assigned to the closest clusters based on distance calculation regarding centroid points. For example, the dataset illustrated in figure is assigned two centroids.

The distance between any object from the dataset to both of the centroid points are calculated and the objects are assigned to the closest cluster represented by the closest centroid point as seen in figure. Then new centroid points of clusters are calculated and objects are assigned to the closest clusters regarding the distance to new centroid points. Recalculation of centroid points and assignment of objects to new clusters goes on till centroids points remain the same as in figure.

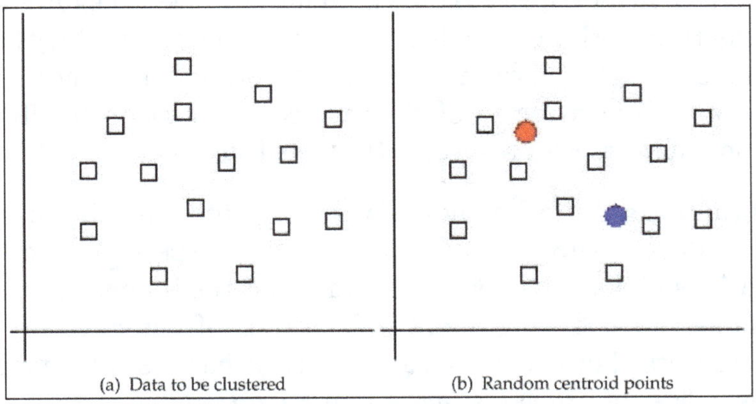

The dataset and two centroid points.

Another method, Self-organizing Map (SOM), is one of the machine-learning techniques widely used in gene clustering. SOM requires a grid structured input that makes it ineffective.

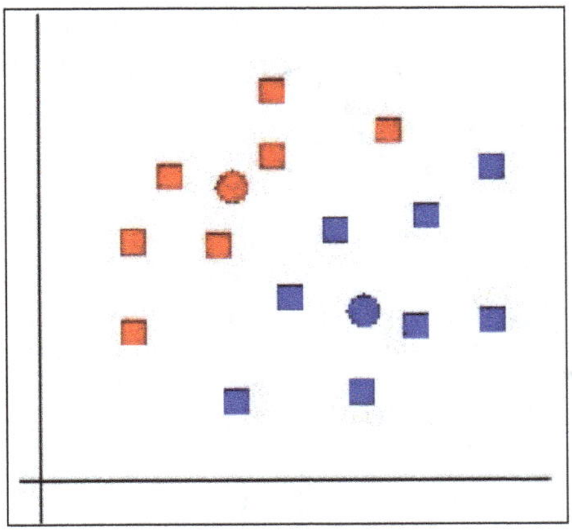

Initial clusters.

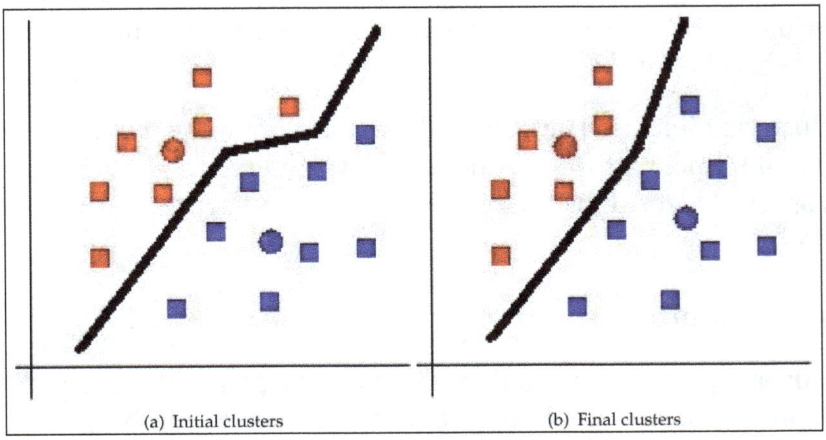

Iteration of K-means.

Hierarchical clustering (HC) algorithms are also widely used and area of two types: agglomerative and divisive. In agglomerative approach objects are all in different clusters and they are merged till they are all in the same cluster as seen in figure. Two important drawbacks of the HC algorithms are that they are not robust and they have high computational complexity. HC algorithms are "greedy" which often means that the final solution is suboptimal due to locally optimal choices being made in initial steps, which turn out to be poor choices with respect to the global solution.

Graph-theoretical clustering techniques exist in which the genomic data are represented by nodes and edges of a graph. Network methods have been applied to identify and characterize various biological interactions. Identification of clusters using networks is often intractable, that is finding an optimal partition of a graph is an NP-hard problem. NP-hard is a class of problems that are at least as hard as NP-complete problems. NP-complete is a class of problems that are in NP and reducible to an NP-complete problem in polynomial time.

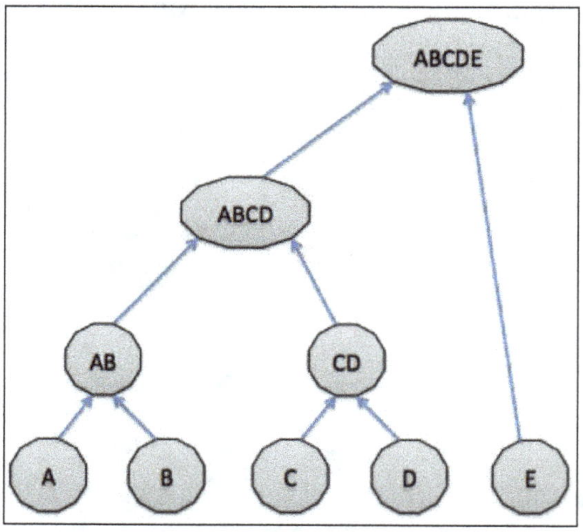

Agglomerative Approach.

Model-based clustering approaches are the ones using probability distributions to predict the distribution of gene expression data. However, gene expression data does not have a unique distribution.

Sub-space clustering (biclustering) methods, which employ the reasoning that one gene may belong to multiple pathways or no pathways. There are also optimization-based algorithms, spectral algorithms, fuzzy algorithms, meta-heuristics used for clustering genomic data.

Ensemble Clustering

Combining diverse partitions from different clustering algorithms may result in high quality and robust clusters, since ensemble approaches such as bagging and boosting used in classification problems have proven to be effective. The fact that the objects have various features makes it difficult to find an optimal clustering of similar objects. In other words, objects may be classified based on different features such as size, color, and age. In that sense, ensemble clustering is a promising heuristic combining results based on different features.

Figure represents a clustering ensemble framework. CAs are clustering algorithms, Ps are partitions generated by them, N is number of clustering algorithms and partitions F_C is the consensus function and CP is the consensus partition.

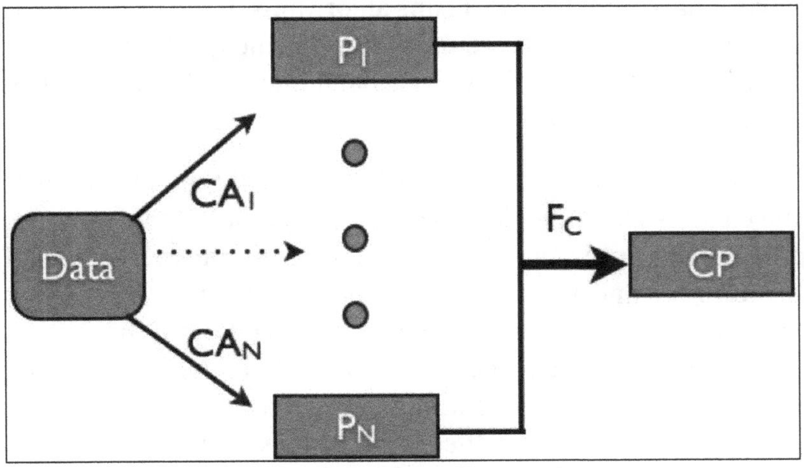

Ensemble clustering framework.

Ensemble clustering requires the following tasks:

- Selection of base clustering algorithms.
- Definition of a consensus function.
- Merging of individual partitions by the chosen consensus function.

Apply an ensemble approach for clustering scale-free graphs. They use metrics based on the neighborhood which uses the adjacency list of each node and considers the nodes as having several common neighbors, the clustering coefficient, and the shortest path betweenness of nodes in the network. The scale-free graph used in the study is from a budding yeast PPI network that contained 15147 interactions between 4741 proteins. It is reported that ensemble clustering can provide improvements in cluster quality for scale-free graphs based upon the preliminary results. Propose an ensemble clustering framework to extract functional modules that are relevant biologically in protein-protein interaction (PPI) networks. Their method attempts to handle the noisy false positive interactions and specific topological interactions present in the network. The method uses graph clustering algorithms, repeated bisections, direct k-way partitioning, and multilevel k-way partitioning, to obtain the base partitions. The method utilizes two topological distance matrices. One of the distance matrices is based on the clustering coefficient, and the other distance matrix is generated using the betweenness measure. The proposed study demonstrates a soft ensemble method such that proteins are allowed to be assigned to more than one cluster. Empirical evaluation of the different ensemble methods in the study shows the superior performance of the proposed ensemble framework.

Fuzzy clustering algorithms are widely used with well-understood properties and benefits in various applications. Nevertheless, there has been very little analysis of using fuzzy clustering algorithms in regards to generating the base partitions in cluster ensembles. Compares hard and fuzzy C-means algorithms in the well-known evidence-accumulation framework of cluster ensembles. In the study, it is observed that the fuzzy C-means approach requires much fewer base partitions for the cluster ensemble to converge, and is more tolerant of outliers in the data.

Propose a fuzzy ensemble clustering approach to address the issue of unclear boundaries between the clusters from the biological and biomedical gene expression data analysis. The approach takes into account their inherent fuzziness. The goal of the study is improving the accuracy and robustness of clustering results. After applying random projections to obtain lower dimensional gene expression data, the method applies the fuzzy K-means algorithm on the low dimensional data to generate multiple fuzzy base clusters. Then, the fuzzy clusters are combined using a similarity matrix where the elements of the matrix are generated by the fuzzy t-norms algorithm, and finally, the fuzzy K-means algorithm is applied to the rows of the similarity matrix to obtain the consensus clustering. It is demonstrated that the proposed ensemble approach is competitive with the other ensemble methods.

High throughput data may be generated by microarray experiments. If the dataset is very large, it is possible to generate an ensemble of clustering solutions, or partition the data so that clustering may be performed on tractable-sized disjoint subsets. The data can then be distributed at different sites, for which a distributed clustering solution with a final merging of partitions is a natural fit. Introduce two new approaches to combining partitions represented by sets of cluster centers. It is stated that these approaches provide a final partition of data that is comparable to the best existing approaches and that the approaches can be 100,000 times faster while using much less memory. The new algorithms are compared with the best existing cluster ensemble approaches that cluster all of the data at once, and a clustering algorithm designed for very large datasets. Fuzzy and hard K-means based clustering algorithms are used for the comparison. It is demonstrated that the centroid-based ensemble merging algorithms presented in the study generated partitions which are as good as the best label vector method, or the method of clustering all the data at once. The proposed algorithms are also more efficient in terms of speed.

Propose evidence accumulation clustering based on dual rooted prim tree cuts (EAC-DC). The proposed algorithm computes the co-association matrix based on a forward algorithm that repeatedly adds edges to Prim's minimum spanning tree (MST) to identify clusters until a satisfying criterion is met. A consensus cluster is then generated from the co-association matrix using spectral partitioning. Here, a MST is a fully connected sub-graph with no cycles and a dual-rooted tree is obtained by finding the union of two sub-trees. They test their approach using the Iris dataset, the Wisconsin breast cancer dataset and synthetic datasets, and presented a comparison of their results with other existing ensemble clustering methods.

Use a cluster ensemble in gene expression analysis. In the proposed ensemble framework, the partitions generated by each individual clustering algorithm are converted into a distance matrix. The distance matrices are then combined to construct a weighted graph. A graph partitioning approach is then used to generate the final set of clusters. It is reported that the ensemble approach yields better results than the best individual approach on both synthetic and yeast gene expression datasets.

Merge multiple partitions using evidence accumulation. Each partition generated by a clustering algorithm is used as a new piece of knowledge, to help uncover the relationships between objects. We adopt their ensemble approach. The core idea behind the ensemble approach here is constructing the co-association matrix by employing a voting mechanism for the partitions generated using individual clustering algorithms. A co-association matrix C is constructed based upon the formulation below, where n_{ij} is the number of times the object pair (i, j) is assigned to the same cluster among the N different partitions:

$$C(i,j) = \frac{n_{ij}}{N}.$$

After constructing the co-association matrix, use single linkage hierarchical clustering to obtain the new cluster tree (dendrogram) and then use a cut-off value corresponding to the maximum life time (difference between merge points where branching starts) on the tree.

They also employ the same ensemble framework using K-means partitions with different parameters. They test their algorithms on ten different datasets, comparing the results with other ensemble clustering methods. They report that their ensemble approach can identify the clusters with arbitrary shapes and sizes, and perform better than the other combination methods.

Implementation

Different set of base clustering algorithms are chosen and implemented on protein and lymphoma datasets. Protein dataset consists of 698 objects (corresponding to protein folds) with 125 attributes. The protein dataset contains 698 proteins from 125 samples. The real clusters correspond to the four classes of protein - folds: α, β, α/β and $\alpha+\beta$ pro tein classes. DLBCL - B is 2 - channel custom cDNA microarray dataset. This is a B cell lymphoma dataset with predefined three subtypes.

The ensemble clustering algorithm uses an array of vectors data structure for each of the file, in order to use the dynamic memory allocation and starts with initializing the file content in the vectors. The algorithm also processes the vectors and generates two temporary matrices with the dimension of maximum vector length. The ensemble clustering algorithm steps are as follows:

Algorithm 1 Ensemble Clustering Algorithm

Require: partitions
Ensure: distance matrix

```
for i = 0 to max(V[n]) do
  for j = 0 to max(V[n]) do
    for k = 0 to n do
      if V[k].elementAt(i) = V[k].elementAt(j)) then
        C[i][j] = C[i][j] + 1/n
      end if
      D[i][j] = 1 − C[i][j]
    end for
  end for
end for
```

Here, n is the number of files, V[n] are the vectors holding the content of each file. *max*-(V[n]) is the length of the longest vector, C[i][j] is the co-association matrix and D[i][j] is the distance matrix. The algorithm iterates through the two dimensional matrix via iand j loop variables inside a nested loop at lines 1 and 2 and for each member of the matrix, all the vectors are processed inside the loop via k loop variable at line 3. The condition of equality for the selected vector with the selected loop variables iand j, causes an increase on the co-association matrix elements at lines 4 and 5. Finally the distance matrix is calculated at line 7. After obtaining the distance matrix, hierarchial clustering with complete linkage is used to generate the dengrogram. The dendrogram is cut at a certain level to obtain consensus partition.

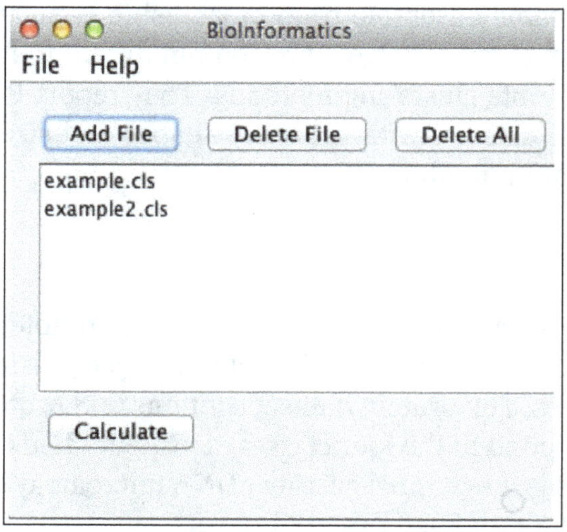

File input interface.

Ensemble approach is coded as a java application which is available upon request. The software allows addition of many partitions to generate the distance matrix of the corresponding ensemble. Files including the partitions can be added by clicking on the "Add File" button as seen in figure. Distance matrix of the ensemble is generated by "Calculate" button.

The output is displayed on a separate screen as demonstrated on figure. The output with csv format can be written into a file by clicking on the "Output CSV" button.

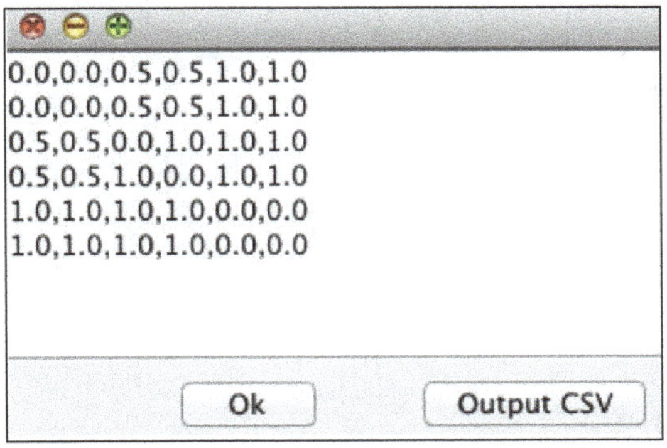

Example Clusters.

Considering two different partitions of a dataset with six objects which are (1, 1, 2, 1, 3, 3) and (2, 2, 2, 1, 3, 3), the algorithm's output is the distance matrix:

$$\begin{pmatrix} 0 & 0 & 0.5 & 0.5 & 1 & 1 \\ 0 & 0 & 0.5 & 0.5 & 1 & 1 \\ 0.5 & 0.5 & 0 & 1 & 1 & 1 \\ 0.5 & 0.5 & 1 & 0 & 1 & 1 \\ 1 & 1 & 1 & 1 & 0 & 0 \\ 1 & 1 & 1 & 1 & 0 & 0 \end{pmatrix}$$

The distance matrix is used in hierarchical clustering with complete linkage and the following dendrogram is generated. The dendrogram is cut at a level to give three clusters. The corresponding partition is (1, 1, 1, 2, 3, 3) which is the same as second partition (2, 2, 2, 1, 3, 3).

We employ hierarchical clustering, K-means and C-means to obtain base partitions. K-means and hierarchical clustering algorithm are implemented using R base package, C-means is implemented using R e1071 package. Silhouette and C-rand indices are utilized to evaluate the performance of individual and ensemble algorithms. Silhouette and C-rand values are calculated using R clusterSim and flexclust packages respectively. Silhouette is an internal measure of compactness and separation of clusters. The silhouette index values are between -1 and 1 representing worst and best values. C-rand is an external measure of agreement between two partitions. C-rand has maximum value of 1 and it can take negative values. The silhouette and C-rand values found by the base and ensemble algorithms are given in Table. Ensemble approach improves clustering result both for the protein and DLBCL-B datasets. Ensemble approach finds better C-rand value, 0.157 than values by K-means and C-means, 0.127 for the protein dataset. Ensemble approach also finds the best C-rand value, 0.135 compared to values generated by individual clustering algorithms, 0.021, 0.063, 0.098. However, the ensemble approach makes S values worse in most cases.

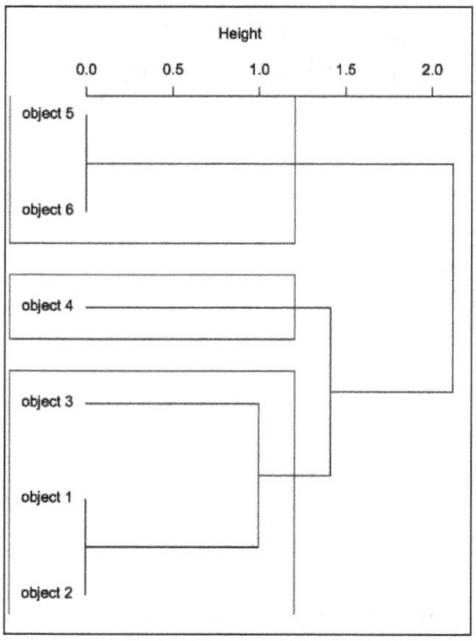

Example Clusters.

Table: Table Index values for base and ensemble algorithms.

Dataset	Method	Num. of clusters	S value	C value
Protein	HC	4	0.344	0.199
	K-means	4	0.379	0.127
	C-means	4	0.379	0.127
	Ensemble	4	0.078	0.157
DLBCL-B	HC	3	-0.034	0.021
	K-means	3	-0.015	0.063
	C-means	2	-0.005	0.098
	Ensemble	3	-0.017	0.135

TYPES OF BIOLOGICAL DATABASES

Biological databases can be roughly divided into two categories:

Primary Databases

- Primary databases are also called as archieval database.
- They are populated with experimentally derived data such as nucleotide sequence, protein sequence or macromolecular structure.

- Experimental results are submitted directly into the database by researchers, and the data are essentially archival in nature.
- Once given a database accession number, the data in primary databases are never changed: They form part of the scientific record.

Examples:

- ENA, GenBank and DDBJ (nucleotide sequence).
- Array Express Archive and GEO (functional genomics data).
- Protein Data Bank (PDB; coordinates of three-dimensional macromolecular structures).

Secondary Databases

- Secondary databases comprise data derived from the results of analysing primary data.
- Secondary databases often draw upon information from numerous sources, including other databases (primary and secondary), controlled vocabularies and the scientific literature.
- They are highly curated, often using a complex combination of computational algorithms and manual analysis and interpretation to derive new knowledge from the public record of science.

Examples:

- InterPro (protein families, motifs and domains).
- UniProt Knowledgebase (sequence and functional information on proteins).
- Ensembl (variation, function, regulation and more layered onto whole genome sequences).

However, many data resources have both primary and secondary characteristics. For example, UniProt accepts primary sequences derived from peptide sequencing experiments. However, UniProt also infers peptide sequences from genomic information, and it provides a wealth of additional information, some derived from automated annotation (TrEMBL), and even more from careful manual analysis (SwissProt).

There are also specialized databases are those that cater to a particular research interest. For example, Flybase, HIV sequence database, and Ribosomal Database Project are databases that specialize in a particular organism or a particular type of data.

SEQUENCE DATABASE

A sequence database is a type of biological database that is composed of a large collection of computerized ("digital") nucleic acid sequences, protein sequences, or other polymer sequences stored on a computer. The UniProt database is an example of a protein sequence database. As of 2013 it contained over 40 million sequences and is growing at an exponential rate. Historically, sequences were published in paper form, but as the number of sequences grew, this storage method became unsustainable.

Sequence databases can be searched using a variety of methods. The most common usage is probably searching for sequences similar to a certain target protein or gene whose sequence is already known to the user. The BLAST program is a popular method of this type.

Records in sequence databases are deposited from a wide range of sources, from individual researchers to large genome sequencing centers. As a result, the sequences themselves, and especially the biological annotations attached to these sequences, may vary in quality. There is much redundancy, as multiple labs may submit numerous sequences that are identical, or nearly identical, to others in the databases.

Many annotations of the sequences are based not on laboratory experiments, but on the results of sequence similarity searches for previously-annotated sequences. Once a sequence has been annotated based on similarity to others, and itself deposited in the database, it can also become the basis for future annotations. This can lead to a *transitive annotation problem* because there may be several such annotation transfers by sequence similarity between a particular database record and actual wet lab experimental information. Therefore, care must be taken when interpreting the annotation data from sequence databases.

The main sequence databases are Genbank and EMBL. Originally they were just sequence collections, but they have grown to store different biological databases heavily interconnected and they provide powerful interfaces to search and browse the stored information.

The sequences submitted to any of those databases are shared between them, so any sequence could be retrieved in the european or the american database. But they differ in the tools to search and browse the data and in some databases that provide extra information to the raw sequences like: mutations, coded proteins, bibliographical references, etc.

These databases are growing at an ever increasing fast pace. In June of 2007 there were 73 million sequences in Genbank and in August of 2015 there were 187 millions.

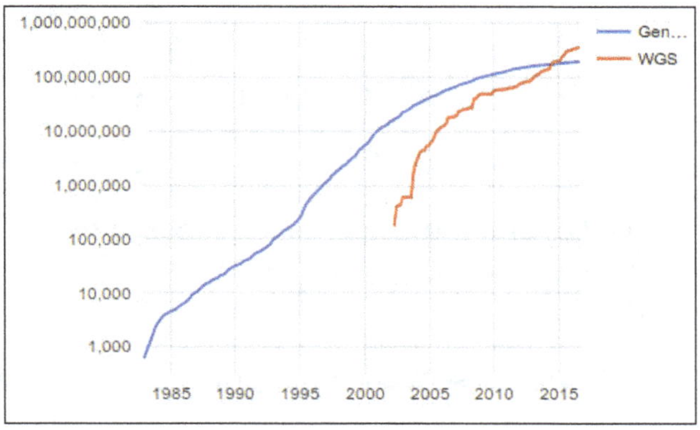

The sequences are split in these databases in different sections to ease the search. Among others, there are sections for mRNAs, publised nucleotide sequences, genomes, and genes.

GENBANK

The GenBank sequence database is an open access, annotated collection of all publicly available nucleotide sequences and their protein translations. This database is produced and maintained by the National Center for Biotechnology Information (NCBI; a part of the National Institutes of Health in the United States) as part of the International Nucleotide Sequence Database Collaboration (INSDC).

GenBank and its collaborators receive sequences produced in laboratories throughout the world from more than 100,000 distinct organisms. The database started in 1982 by Walter Goad and Los Alamos National Laboratory. GenBank has become an important database for research in biological fields and has grown in recent years at an exponential rate by doubling roughly every 18 months.

GenBank is built by direct submissions from individual laboratories, as well as from bulk submissions from large-scale sequencing centers.

Only original sequences can be submitted to GenBank. Direct submissions are made to GenBank using BankIt, which is a Web-based form, or the stand-alone submission program, Sequin. Upon receipt of a sequence submission, the GenBank staff examines the originality of the data and assigns an accession number to the sequence and performs quality assurance checks. The submissions are then released to the public database, where the entries are retrievable by Entrez or downloadable by FTP. Bulk submissions of Expressed Sequence Tag (EST), Sequence-tagged site (STS), Genome Survey Sequence (GSS), and High-Throughput Genome Sequence (HTGS) data are most often submitted by large-scale sequencing centers. The GenBank direct submissions group also processes complete microbial genome sequences.

Access to GenBank

There are several ways to search and retrieve data from GenBank:

- Search GenBank for sequence identifiers and annotations with Entrez Nucleotide.
- Search and align GenBank sequences to a query sequence using BLAST (Basic Local Alignment Search Tool). BLAST searches CoreNucleotide, dbEST, and dbGSS independently.
- Search, link, and download sequences programatically using NCBI e-utilities.

GenBank Data Usage

The GenBank database is designed to provide and encourage access within the scientific community to the most up-to-date and comprehensive DNA sequence information. Therefore, NCBI places no restrictions on the use or distribution of the GenBank data. However, some submitters may claim patent, copyright, or other intellectual property rights in all or a portion of the data they have submitted. NCBI is not in a position to assess the validity of such claims, and therefore cannot provide comment or unrestricted permission concerning the use, copying, or distribution of the information contained in GenBank.

ENTREZ PROTEIN

Entrez is a molecular biology database system that provides integrated access to nucleotide and protein sequence data, gene-centered and genomic mapping information, 3D structure data, PubMed MEDLINE, and more. The system is produced by the National Center for Biotechnology Information (NCBI) and is available via the Internet.

Entrez Databases and Retrieval System

Entrez covers over 20 databases including the complete protein sequence data from PIR-International, PRF, Swiss-Prot, and PDB and nucleotide sequence data from GenBank that includes information from EMBL and DDBJ.

The Entrez retrieval system uses an intuitive user interface for rapidly searching sequence and bibliographic data. A unique feature of the system is its use of precomputed similarity searches for each record to create links to "neighbors" or related records in other Entrez databases. These links facilitate integrated access across the various databases. An Entrez global query provides search capability for a subset of Entrez databases at one time. Results may be viewed in various formats inlcuding FlatFile, FASTA, XML, and others. A graphical interface provides easy visualization of complete genomes or chromosomes, as well as biological annotation on individual sequences. Entrez also allows Batch downloads of large search results.

Features

The Entrez front page provides, by default, access to the global query. All databases indexed by Entrez can be searched via a single query string, supporting boolean operators and search term tags to limit parts of the search statement to particular fields. This returns a unified results page, that shows the number of hits for the search in each of the databases, which are also links to actual search results for that particular database.

Entrez also provides a similar interface for searching each particular database and for refining search results. The Limits feature allows the user to narrow a search a web forms interface. The History feature gives a numbered list of recently performed queries. Results of previous queries can be referred to by number and combined via boolean operators. Search results can be saved temporarily in a Clipboard. Users with a MyNCBI account can save queries indefinitely and also choose to have updates with new search results e-mailed for saved queries of most databases. It is widely used in the field of biotechnology as a reference tool for students and professionals alike.

UNIPROT

UniProt is a freely accessible database of protein sequence and functional information, many entries being derived from genome sequencing projects. It contains a large amount of information about the biological function of proteins derived from the research literature.

UniProt Consortium

The UniProt consortium comprises the European Bioinformatics Institute (EBI), the Swiss Institute of Bioinformatics (SIB), and the Protein Information Resource (PIR). EBI, located at the Wellcome Trust Genome Campus in Hinxton, UK, hosts a large resource of bioinformatics databases and services. SIB, located in Geneva, Switzerland, maintains the ExPASy (Expert Protein Analysis System) servers that are a central resource for proteomics tools and databases. PIR, hosted by the National Biomedical Research Foundation (NBRF) at the Georgetown University Medical Center in Washington, DC, US, is heir to the oldest protein sequence database, Margaret Dayhoff's Atlas of Protein Sequence and Structure, first published in 1965. In 2002, EBI, SIB, and PIR joined forces as the UniProt consortium.

Roots of UniProt Databases

Each consortium member is heavily involved in protein database maintenance and annotation. Until recently, EBI and SIB together produced the Swiss-Prot and TrEMBL databases, while PIR produced the Protein Sequence Database (PIR-PSD). These databases coexisted with differing protein sequence coverage and annotation priorities.

Swiss-Prot was created in 1986 by Amos Bairoch during his PhD and developed by the Swiss Institute of Bioinformatics and subsequently developed by Rolf Apweiler at the European Bioinformatics Institute. Swiss-Prot aimed to provide reliable protein sequences associated with a high level of annotation (such as the description of the function of a protein, its domain structure, post-translational modifications, variants, etc.), a minimal level of redundancy and high level of integration with other databases. Recognizing that sequence data were being generated at a pace exceeding Swiss-Prot's ability to keep up, TrEMBL (Translated EMBL Nucleotide Sequence Data Library) was created to provide automated annotations for those proteins not in Swiss-Prot. Meanwhile, PIR maintained the PIR-PSD and related databases, including iProClass, a database of protein sequences and curated families. The consortium members pooled their overlapping resources and expertise, and launched UniProt in December 2003.

Organization of UniProt Databases

UniProt provides four core databases: UniProtKB (with sub-parts Swiss-Prot and TrEMBL), UniParc, UniRef.

UniProtKB

UniProt Knowledgebase (UniProtKB) is a protein database partially curated by experts, consisting of two sections: UniProtKB/Swiss-Prot (containing reviewed, manually annotated entries) and UniProtKB/TrEMBL (containing unreviewed, automatically annotated entries). As of 19 March 2014, release "2014_03" of UniProtKB/Swiss-Prot contains 542,782 sequence entries (comprising 193,019,802 amino acids abstracted from 226,896 references) and release "2014_03" of UniProtKB/TrEMBL contains 54,247,468 sequence entries (comprising 17,207,833,179 amino acids).

UniProtKB/Swiss-Prot

UniProtKB/Swiss-Prot is a manually annotated, non-redundant protein sequence database. It combines information extracted from scientific literature and biocurator-evaluated computational analysis. The aim of UniProtKB/Swiss-Prot is to provide all known relevant information about a particular protein. Annotation is regularly reviewed to keep up with current scientific findings. The manual annotation of an entry involves detailed analysis of the protein sequence and of the scientific literature.

Sequences from the same gene and the same species are merged into the same database entry. Differences between sequences are identified, and their cause documented (for example alternative splicing, natural variation, incorrect initiation sites, incorrect exon boundaries, frameshifts, unidentified conflicts). A range of sequence analysis tools is used in the annotation of UniProtKB/Swiss-Prot entries. Computer-predictions are manually evaluated, and relevant results selected for inclusion in the entry. These predictions include post-translational modifications, transmembrane domains and topology, signal peptides, domain identification, and protein family classification.

Relevant publications are identified by searching databases such as PubMed. The full text of each paper is read, and information is extracted and added to the entry. Annotation arising from the scientific literature includes, but is not limited to:

- Protein and gene names.
- Function.
- Enzyme-specific information such as catalytic activity, cofactors and catalytic residues.
- Subcellular location.
- Protein-protein interactions.
- Pattern of expression.
- Locations and roles of significant domains and sites.
- Ion, substrate and cofactor binding sites.
- Protein variant forms produced by natural genetic variation, RNA editing, alternative splicing, proteolytic processing, and post-translational modification.

Annotated entries undergo quality assurance before inclusion into UniProtKB/Swiss-Prot. When new data becomes available, entries are updated.

UniProtKB/TrEMBL

UniProtKB/TrEMBL contains high-quality computationally analyzed records, which are enriched with automatic annotation. It was introduced in response to increased dataflow resulting from genome projects, as the time- and labour-consuming manual annotation process of UniProtKB/Swiss-Prot could not be broadened to include all available protein sequences. The translations of annotated coding sequences in the EMBL-Bank/GenBank/DDBJ nucleotide sequence database are automatically processed and entered in UniProtKB/TrEMBL. UniProtKB/TrEMBL also contains sequences from PDB, and from gene prediction, including Ensembl, RefSeq and CCDS.

UniParc

UniProt Archive (UniParc) is a comprehensive and non-redundant database, which contains all the protein sequences from the main, publicly available protein sequence databases. Proteins may exist in several different source databases, and in multiple copies in the same database. In order to avoid redundancy, UniParc stores each unique sequence only once. Identical sequences are merged, regardless of whether they are from the same or different species. Each sequence is given a stable and unique identifier (UPI), making it possible to identify the same protein from different source databases. UniParc contains only protein sequences, with no annotation. Database cross-references in UniParc entries allow further information about the protein to be retrieved from the source databases. When sequences in the source databases change, these changes are tracked by UniParc and history of all changes is archived.

Source Databases

Currently UniParc contains protein sequences from the following publicly available databases:

- INSDC EMBL-Bank/DDBJ/GenBank nucleotide sequence databases.
- Ensembl.
- European Patent Office (EPO).
- FlyBase: the primary repository of genetic and molecular data for the insect family. Drosophilidae (FlyBase).
- H-Invitational Database (H-Inv).
- International Protein Index (IPI).
- Japan Patent Office (JPO).
- Protein Information Resource (PIR-PSD).
- Protein Data Bank (PDB).
- Protein Research Foundation (PRF).
- RefSeq.
- Saccharomyces Genome Database (SGD).
- The Arabidopsis Information Resource (TAIR).
- TROME.
- US Patent Office (USPTO).
- UniProtKB/Swiss-Prot, UniProtKB/Swiss-Prot protein isoforms, UniProtKB/TrEMBL.

- Vertebrate and Genome Annotation Database (VEGA).
- WormBase.

UniRef

The UniProt Reference Clusters (UniRef) consist of three databases of clustered sets of protein sequences from UniProtKB and selected UniParc records. The UniRef100 database combines identical sequences and sequence fragments (from any organism) into a single UniRef entry. The sequence of a representative protein, the accession numbers of all the merged entries and links to the corresponding UniProtKB and UniParc records are displayed. UniRef100 sequences are clustered using the CD-HIT algorithm to build UniRef90 and UniRef50. Each cluster is composed of sequences that have at least 90% or 50% sequence identity, respectively, to the longest sequence. Clustering sequences significantly reduces database size, enabling faster sequence searches.

PROTEIN INFORMATION RESOURCE

The Protein Information Resource (PIR) produces the largest, most comprehensive, annotated protein sequence database in the public domain, the PIR-International Protein Sequence Database, in collaboration with the Munich Information Center for Protein Sequences (MIPS) and the Japan International Protein Sequence Database (JIPID).

The accelerating pace of genome sequencing projects has greatly increased the volume and complexity of available molecular data. To realize the fullest possible value from the data and to gain a better understanding of the genome, databases and the computational tools for analyzing them are required to allow biologically relevant features in the sequences to be identified and to provide insight on their structure and function. For over 30 years, the Protein Information Resource (PIR) has been providing the scientific community with databases and tools for the organization and analysis of protein sequence data. Together with MIPS and JIPID, we have undertaken a major restructuring to meet the challenges presented by the rapid growth of largely uncharacterized sequence data and the opportunities provided by the nearly universal access of scientists to the resources available on the WWW. Among the key developments are complete protein family organization for the PIR-International Protein Sequence Database (PSD) and integrated WWW interfaces for user-friendly sequence analysis, database searching and information retrieval.

PIR-International Protein Databases

PIR, MIPS and JIPID constitute the PIR-International consortium that maintains the PIR-International Protein Sequence Database (PSD), the largest publicly distributed and freely available protein sequence database. The database has the following distinguishing features:

- It is a comprehensive, annotated, and non-redundant protein sequence database, containing over 142 000 sequences as of September 1999. Included are sequences from the completely sequenced genomes of 16 prokaryotes, six archaebacteria, 17 viruses and phages, >100 eukaryote organelles and Saccharomyces cerevisiae.
- The collection is well organized with >99% of entries classified by protein family and >57% classified by protein superfamily.
- PSD annotation includes concurrent cross-references to other sequence, structure, genomic and citation databases, including the public nucleic acid sequence databases ENTREZ, MEDLINE, PDB, GDB, OMIM, FlyBase, MIPS/Yeast, SGD/Yeast, MIPS/Arabidopsis and TIGR. Where these databases are publicly and freely accessible and provide suitable WWW access, the cross-references presented on the PIR WWW site are hot-linked so that searchers can consult the most current data.
- The PIR is the only sequence database to provide context cross-references between its own database entries. These cross-references assist searchers in exploring relationships such as subunit associations in molecular complexes, enzyme–substrate interactions, activation and regulation cascades, as well as in browsing entries with shared features and annotations.
- Interim updates are made publicly available on a weekly basis, and full releases have been published quarterly since 1984.

In addition to the PSD, PIR-International distributes or provides WWW access to other sequence and auxiliary databases, briefly described below, and maintains several internal data collections used for sequence annotation and integrity checks.

- PATCHX is a non-redundant database assembled by MIPS of publicly available protein sequences not yet in the PIR-International PSD. PIR+PATCHX, a combination of the PSD and PATCHX containing ~300 000 sequences available for similarity searches, is the most complete non-redundant collection of protein sequences available in the public domain.
- ARCHIVE is a database of protein sequences as originally reported in a publication or submission, the only such collection of 'as published' unmerged sequences.
- NRL_3D sequence-structure database is produced from sequence and annotation in the Protein Data Bank (PDB) of three-dimensional structures.

- FAMBASE is a collection of representative sequences from each protein family that can be used in a similarity search to reduce search time and improve sensitivity for identifying distant families.
- PIR-ALN is a curated database of sequence alignments of superfamilies, families and homology domains, with annotation information derived from PSD and consensus patterns calculated from the alignments.
- RESID is a database of post-translational modifications with descriptive, chemical, structural and bibliographic information based on feature information in the PSD.
- ProClass is a protein family database that organizes non-redundant PIR-International PSD and SWISS-PROT sequences according to PIR superfamilies and PROSITE patterns.
- ProtFam is a curated database of homology clusters with automatically generated multiple sequence alignments for families, superfamilies and homology domains.

To support both data management and data mining and assist knowledge discovery, the PIR databases are being migrated to an object-relational database management system. A three-tier network computing, architecture provides a framework for distributed object computing and Java-based WWW interfaces connect with the database server for database query and update tasks.

Superfamily and Family Classification

The pioneering work of Margaret Dayhoff on protein classification based on the superfamily concept was refined by PIR-International to assist database organization and molecular evolution studies. Central to the organization and annotation of the PIR-International sequence and auxiliary databases are the protein family relationships which are structured at three levels: (i) Superfamilies and families (for full-length sequence similarity), (ii) Homology domains (for local functional or structural units), and (iii) Motifs (for functional or structural sites). PIR-International has maintained the highest classification rate and provided the most comprehensive classification and alignments of proteins among all major public domain databases.

To deal efficiently with the many new sequences from genome sequencing projects, procedures for family and superfamily classification have been automated. Over 99% of sequences are routinely classified shortly after entry into the database into protein families of sequences that are at least 45% identical. Subsequently, entries are further clustered into regular superfamilies of sequences that share end-to-end homology (but may be rather distantly related) and also into domain superfamilies of proteins sharing at least one common homology domain. There are currently >76 000 sequences in >8900 superfamilies, and 30 000 entries with 370 recognized homology domains in the PSD. Corresponding to the classification are 1500 superfamily, 2100 family and 400 domain alignments in the PIR-ALN database, and 15 000 family and 4500 superfamily alignments in the MIPS ProtFam database. Also available from PIR is the ProClass protein family database containing 92 000 classified entries as well as 1300 motif alignments of ~44 000 PIR-International PSD entries.

PIR Search and Analysis System

The PIR search and analysis system provides search engines of three types: (i) Interactive text-based search engines, which allow Boolean queries of text fields; (ii) Standard sequence similarity search engines, including Peptide Match, Pattern Match, BLAST, FASTA, Pairwise Alignment and Multiple Alignment; and (iii) Advanced search engines that combine sequence similarity and annotation searches or evaluate gene family relationships, including Annotation-Sorted Similarity Search, Domain Search, Global and Domain Similarity Search and GeneFIND. Sequence searching can be performed against the PSD, NRL_3D, PATCHX, FAMBASE, ProClass and the combined PSD+PATCHX collections. Text and entry searching is provided for the PSD, NRL_3D, PIR-ALN and RESID databases.

Sequence Search and Alignment

BLAST and FASTA searches for sequence similarity are available for all sequence databases. The output of these search engines employs a graphical interface showing location of hits within the query sequence and full-length alignments generated by SSEARCH. Multiple or pairwise alignments of PSD or user-supplied sequences can be done using CLUSTALW or SSEARCH. PIR pattern or peptide matching programs can: (i) Match a query sequence against a database of regular expressions (i.e., patterns); (ii) Search a user-specified regular expression against a sequence database; or (iii) Find an exact match for a user-specified peptide sequence in one of the sequence databases, including the ARCHIVE database of 'as published' sequences.

PIR Similarity Search System

Combining sequence and annotation search, the Annotation-Sorted Similarity Search facility displays BLAST or FASTA matches along with the user-selected annotation (superfamily, family, species, taxonomic group, keyword or all five) in the annotation-sorted order. The matched entries can be selected for multiple alignments against the query sequence using CLUSTALW and displayed using MView.

The Domain Similarity Search engine uses FASTA to search against domain sequences compiled from the PIR-International PSD, and displays the PSD entry and domain annotation with a graphical representation of the matched region with links to domain alignments in PIR-ALN. The Global and Domain Similarity Search uses BLAST to search the PSD for global similarity and FASTA to search the domain sequence collection for local similarity. The results are ranked by the global score and show the extent of matches at both the global and domain levels. Any combination of complete sequences and domains can be selected and viewed in a multiple alignment.

The PIR Integrated Environment for Sequence Analysis provides an integrated environment for all above protein analysis tasks, including sequence similarity search, pattern and peptide match, multiple sequence alignment and advanced PIR similarity searches, as well as for entry retrieval by unique superfamily, family, title, species, taxonomic group, domains or keywords.

GeneFIND provides protein family classification and information retrieval by combining several search/alignment tools and the ProClass database in a multi-level filter system, including the MOTIFIND neural networks, BLAST search, SSEARCH sequence alignment, motif pattern matching, hidden Markov motif modeling and CLUSTALW multiple motif alignment.

PROTEIN STRUCTURE DATABASE

A protein structure database is a database that is modeled around the various experimentally determined protein structures. The aim of most protein structure databases is to organize and annotate the protein structures, providing the biological community access to the experimental data in a useful way. Data included in protein structure databases often includes three-dimensional coordinates as well as experimental information, such as unit cell dimensions and angles for x-ray crystallography determined structures. Though most instances, in this case either proteins or a specific structure determinations of a protein, also contain sequence information and some databases even provide means for performing sequence based queries, the primary attribute of a structure database is structural information, whereas sequence databases focus on sequence information, and contain no structural information for the majority of entries. Protein structure databases are critical for many efforts in computational biology such as structure based drug design, both in developing the computational methods used and in providing a large experimental dataset used by some methods to provide insights about the function of a protein.

PROTEIN DATA BANK

The Protein Data Bank (PDB) is a database for the three-dimensional structural data of large biological molecules, such as proteins and nucleic acids. The data, typically obtained by X-ray crystallography, NMR spectroscopy, or, increasingly, cryo-electron microscopy, and submitted by biologists and biochemists from around the world, are freely accessible on the Internet via the websites of its member organisations (PDBe, PDBj, RCSB, and BMRB). The PDB is overseen by an organization called the Worldwide Protein Data Bank, wwPDB.

The PDB is a key in areas of structural biology, such as structural genomics. Most major scientific journals, and some funding agencies, now require scientists to submit their structure data to the PDB. Many other databases use protein structures deposited in the PDB. For example, SCOP and CATH classify protein structures, while PDBsum provides a graphic overview of PDB entries using information from other sources, such as Gene ontology.

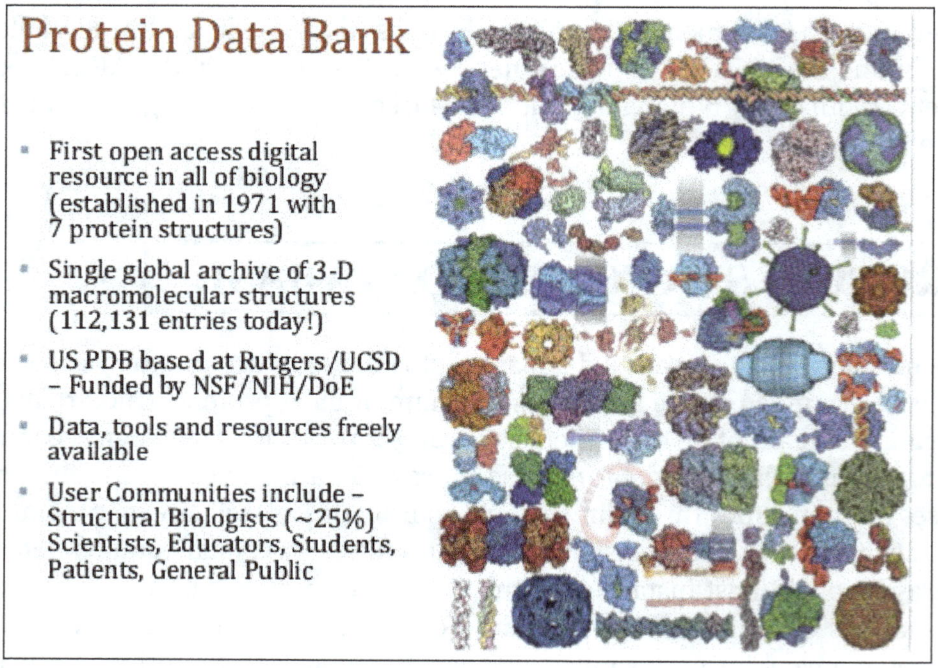

Protein structures from the PDB.

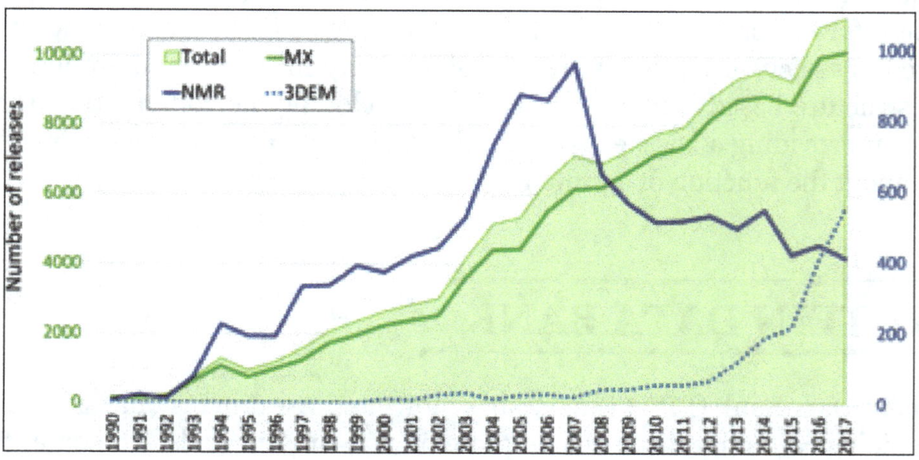

Rate of protein structure determination by method and year.

The PDB database is updated weekly (UTC+0 Wednesday), along with its holdings list. As of 4 July 2019, the PDB comprised:

Experimental Method	Proteins	Nucleic Acids	Protein/Nucleic Acid complexes	Other	Total
X-ray diffraction	128397	2029	6628	4	137058
NMR	11112	1290	260	8	12670
Electron microscopy	2524	32	885	0	3441
Hybrid	141	5	3	1	150
Other	259	4	6	13	282
Total:	142433	3360	7782	26	153601

- 126,949 structures in the PDB have a structure factor file.
- 10,022 structures have an NMR restraint file.
- 3,774 structures in the PDB have a chemical shifts file.
- 3,484 structures in the PDB have a 3DEM map file deposited in EM Data Bank.

Most structures are determined by X-ray diffraction, but about 10% of structures are determined by protein NMR. When using X-ray diffraction, approximations of the coordinates of the atoms of the protein are obtained, whereas using NMR, the distance between pairs of atoms of the protein is estimated. The final conformation of the protein is obtained from NMR by solving a distance geometry problem. A few proteins are determined by cryo-electron microscopy. Clicking on the numbers in the linked external table displays examples of structures determined by that method.

For PDB structures determined by X-ray diffraction that have a structure factor file, their electron density map may be viewed. The data of such structures is stored on the "electron density server".

Historically, the number of structures in the PDB has grown at an approximately exponential rate, with 100 registered structures in 1982, 1,000 structures in 1993, 10,000 in 1999, and 100,000 in 2014. Since 2007, the rate of accumulation of new protein structures appears to have plateaued.

File Format

The file format initially used by the PDB was called the PDB file format. The original format was restricted by the width of computer punch cards to 80 characters per line. Around 1996, the "macromolecular Crystallographic Information file" format, mmCIF, which is an extension of the CIF format was phased in. As of 2013 mmCIF is the master format for the PDB archive. In 2019, the wwPDB announced that depositions for crystallographic methods would only be accepted in mmCIF format.

An XML version of PDB, called PDBML, was described in 2005. The structure files can be downloaded in any of these three formats, though an increasing number of structures do not fit the legacy PDB format. Individual files are easily downloaded into graphics packages from Internet URLs:

- For PDB format files, use, e.g.,

  `http://www.pdb.org/pdb/files/4hhb.pdb.gz` or

  `http://pdbe.org/download/4hhb`

- For PDBML (XML) files, use, e.g.,

  `http://www.pdb.org/pdb/files/4hhb.xml.gz` or

  `http://pdbe.org/pdbml/4hhb`

The "4hhb" is the PDB identifier. Each structure published in PDB receives a four-character alphanumeric identifier, its PDB ID. (This is not a unique identifier for biomolecules, because several structures for the same molecule—in different environments or conformations—may be contained in PDB with different PDB IDs).

Viewing the Data

The structure files may be viewed using one of several free and open source computer programs, including Jmol, Pymol, VMD, and Rasmol. Other non-free, shareware programs include ICM-Browser, MDL Chime, UCSF Chimera, Swiss-PDB Viewer, StarBiochem (a Java-based interactive molecular viewer with integrated search of protein databank), Sirius, and VisProt3DS (a tool for Protein Visualization in 3D stereoscopic view in anaglyth and other modes), and Discovery Studio. The RCSB PDB website contains an extensive list of both free and commercial molecule visualization programs and web browser plugins.

ONLINE MENDELIAN INHERITANCE IN MAN

Online Mendelian Inheritance in Man (OMIM) is a continuously updated catalog of human genes and genetic disorders and traits, with a particular focus on the gene-phenotype relationship. As of 28 June 2019, approximately 9,000 of the over 25,000 entries in OMIM represented phenotypes; the rest represented genes, many of which were related to known phenotypes.

Collection Process and Use

The content of MIM/OMIM is based on selection and review of the published peer-reviewed biomedical literature. Updating of content is performed by a team of science writers and curators under the direction of Dr. Ada Hamosh at the McKusick-Nathans Institute of Genetic Medicine of Johns Hopkins University. While OMIM is freely available to the public, it is designed for use primarily by physicians and other health care professionals concerned with genetic disorders, by genetics researchers, and by advanced students in science and medicine.

The database may be used as a resource for locating literature relevant to inherited conditions, and its numbering system is widely used in the medical literature to provide a unified index for genetic diseases.

MIM Classification System

MIM Numbers

Each OMIM entry is given a unique six-digit identifier as summarized below:

- 100000–299999: Autosomal loci or phenotypes.
- 300000–399999: X-linked loci or phenotypes.
- 400000–499999: Y-linked loci or phenotypes.
- 500000–599999: Mitochondrial loci or phenotypes.
- 600000 and above: Autosomal loci or phenotypes.

In cases of allelic heterogeneity, the MIM number of the entry is followed by a decimal point and a unique 4-digit number specifying the variant. For example, allelic variants in the HBB gene (141900) are numbered 141900.0001 through 141900.0538.

Because OMIM has responsibility for the classification and naming of genetic disorders, these numbers are stable identifiers of the disorders.

Symbols Preceding MIM Numbers

Symbols preceding MIM numbers indicate the entry category:

- An asterisk (*) before an entry number indicates a gene.
- A number symbol (#) before an entry number indicates that it is a descriptive entry, usually of a phenotype, and does not represent a unique locus. The reason for the use of the number symbol is given in the first paragraph of the entry. Discussion of any gene(s) related to the phenotype resides in another entry (or entries) as described in the first paragraph.
- A plus sign (+) before an entry number indicates that the entry contains the description of a gene of known sequence and a phenotype.

- A percent sign (%) before an entry number indicates that the entry describes a confirmed Mendelian phenotype or phenotypic locus for which the underlying molecular basis is not known.
- No symbol before an entry number generally indicates a description of a phenotype for which the Mendelian basis, although suspected, has not been clearly established or that the separateness of this phenotype from that in another entry is unclear.
- A caret (^) before an entry number means the entry no longer exists because it was removed from the database or moved to another entry as indicated.

KYOTO ENCYCLOPEDIA OF GENES AND GENOMES

KEGG (Kyoto Encyclopedia of Genes and Genomes) is a collection of databases dealing with genomes, biological pathways, diseases, drugs, and chemical substances. KEGG is utilized for bioinformatics research and education, including data analysis in genomics, metagenomics, metabolomics and other omics studies, modeling and simulation in systems biology, and translational research in drug development.

The KEGG database project was initiated in 1995 by Minoru Kanehisa, Professor at the Institute for Chemical Research, Kyoto University, under the then ongoing Japanese Human Genome Program. Foreseeing the need for a computerized resource that can be used for biological interpretation of genome sequence data, he started developing the KEGG PATHWAY database. It is a collection of manually drawn KEGG pathway maps representing experimental knowledge on metabolism and various other functions of the cell and the organism. Each pathway map contains a network of molecular interactions and reactions and is designed to link genes in the genome to gene products (mostly proteins) in the pathway. This has enabled the analysis called KEGG pathway mapping, whereby the gene content in the genome is compared with the KEGG PATHWAY database to examine which pathways and associated functions are likely to be encoded in the genome.

According to the developers, KEGG is a "computer representation" of the biological system. It integrates building blocks and wiring diagrams of the system – more specifically, genetic building blocks of genes and proteins, chemical building blocks of small molecules and reactions, and wiring diagrams of molecular interaction and reaction networks. This concept is realized in the following databases of KEGG, which are categorized into systems, genomic, chemical, and health information.

- Systems information:
  - Pathway: Pathway maps for cellular and organismal functions.

- Module: Modules or functional units of genes.
- Brite: Hierarchical classifications of biological entities.

- Genomic information:
  - Genome: Complete genomes.
  - Genes: Genes and proteins in the complete genomes.
  - Orthology: Ortholog groups of genes in the complete genomes.
- Chemical information:
  - Compound, Glycan: Chemical compounds and glycans.
  - Reaction, Rpair, Rclass: Chemical reactions.
  - Enzyme: Enzyme nomenclature.
- Health information:
  - Disease: Human diseases.
  - Drug: Approved drugs.
  - Environ: Crude drugs and health-related substances.

Databases

Systems Information

The KEGG PATHWAY database, the wiring diagram database, is the core of the KEGG resource. It is a collection of pathway maps integrating many entities including genes, proteins, RNAs, chemical compounds, glycans, and chemical reactions, as well as disease genes and drug targets, which are stored as individual entries in the other databases of KEGG. The pathway maps are classified into the following sections:

- Metabolism.
- Genetic information processing (transcription, translation, replication and repair, etc.).
- Environmental information processing (membrane transport, signal transduction, etc.).
- Cellular processes (cell growth, cell death, cell membrane functions, etc.).
- Organismal systems (immune system, endocrine system, nervous system, etc.).

- Human diseases.
- Drug development.

The metabolism section contains aesthetically drawn global maps showing an overall picture of metabolism, in addition to regular metabolic pathway maps. The low-resolution global maps can be used, for example, to compare metabolic capacities of different organisms in genomics studies and different environmental samples in metagenomics studies. In contrast, KEGG modules in the KEGG MODULE database are higher-resolution, localized wiring diagrams, representing tighter functional units within a pathway map, such as subpathways conserved among specific organism groups and molecular complexes. KEGG modules are defined as characteristic gene sets that can be linked to specific metabolic capacities and other phenotypic features, so that they can be used for automatic interpretation of genome and metagenome data.

Another database that supplements KEGG PATHWAY is the KEGG BRITE database. It is an ontology database containing hierarchical classifications of various entities including genes, proteins, organisms, diseases, drugs, and chemical compounds. While KEGG PATHWAY is limited to molecular interactions and reactions of these entities, KEGG BRITE incorporates many different types of relationships.

Genomic Information

Several months after the KEGG project was initiated in 1995, the first report of the completely sequenced bacterial genome was published. Since then all published complete genomes are accumulated in KEGG for both eukaryotes and prokaryotes. The KEGG GENES database contains gene/protein-level information and the KEGG GENOME database contains organism-level information for these genomes. The KEGG GENES database consists of gene sets for the complete genomes, and genes in each set are given annotations in the form of establishing correspondences to the wiring diagrams of KEGG pathway maps, KEGG modules, and BRITE hierarchies.

These correspondences are made using the concept of orthologs. The KEGG pathway maps are drawn based on experimental evidence in specific organisms but they are designed to be applicable to other organisms as well, because different organisms, such as human and mouse, often share identical pathways consisting of functionally identical genes, called orthologous genes or orthologs. All the genes in the KEGG GENES database are being grouped into such orthologs in the KEGG ORTHOLOGY (KO) database. Because the nodes (gene products) of KEGG pathway maps, as well as KEGG modules and BRITE hierarchies, are given KO identifiers, the correspondences are established once genes in the genome are annotated with KO identifiers by the genome annotation procedure in KEGG.

Chemical Information

The KEGG metabolic pathway maps are drawn to represent the dual aspects of the metabolic network: the genomic network of how genome-encoded enzymes are connected to catalyze consecutive reactions and the chemical network of how chemical structures of substrates and products are transformed by these reactions. A set of enzyme genes in the genome will identify enzyme relation networks when superimposed on the KEGG pathway maps, which in turn characterize chemical structure transformation networks allowing interpretation of biosynthetic and biodegradation potentials of the organism. Alternatively, a set of metabolites identified in the metabolome will lead to the understanding of enzymatic pathways and enzyme genes involved.

The databases in the chemical information category, which are collectively called KEGG LIGAND, are organized by capturing knowledge of the chemical network. In the beginning of the KEGG project, KEGG LIGAND consisted of three databases: KEGG COMPOUND for chemical compounds, KEGG REACTION for chemical reactions, and KEGG ENZYME for reactions in the enzyme nomenclature. Currently, there are additional databases: KEGG GLYCAN for glycans and two auxiliary reaction databases called RPAIR (reactant pair alignments) and RCLASS (reaction class). KEGG COMPOUND has also been expanded to contain various compounds such as xenobiotics, in addition to metabolites.

Health Information

In KEGG, diseases are viewed as perturbed states of the biological system caused by perturbants of genetic factors and environmental factors, and drugs are viewed as different types of perturbants. The KEGG PATHWAY database includes not only the normal states but also the perturbed states of the biological systems. However, disease pathway maps cannot be drawn for most diseases because molecular mechanisms are not well understood. An alternative approach is taken in the KEGG DISEASE database, which simply catalogs known genetic factors and environmental factors of diseases. These catalogs may eventually lead to more complete wiring diagrams of diseases.

The KEGG DRUG database contains active ingredients of approved drugs in Japan, the USA, and Europe. They are distinguished by chemical structures and chemical components and associated with target molecules, metabolizing enzymes, and other molecular interaction network information in the KEGG pathway maps and the BRITE hierarchies. This enables an integrated analysis of drug interactions with genomic information. Crude drugs and other health-related substances, which are outside the category of approved drugs, are stored in the KEGG ENVIRON database. The databases in the health information category are collectively called KEGG MEDICUS, which also includes package inserts of all marketed drugs in Japan.

GENE DISEASE DATABASE

In bioinformatics, a Gene Disease Database is a systematized collection of data, typically structured to model aspects of reality, in a way to comprehend the underlying mechanisms of complex diseases, by understanding multiple composite interactions between phenotype-genotype relationships and gene-disease mechanisms. Gene Disease Databases integrate human gene-disease associations from various expert curated databases and text-mining derived associations including Mendelian, complex and environmental diseases.

Experts in different areas of biology and bioinformatics have been trying to comprehend the molecular mechanisms of diseases to design preventive and therapeutic strategies for a long time. For some illnesses, it has become apparent that it is the right amount of animosity is made for not enough to obtain an index of the disease-related genes but to uncover how disruptions of molecular grids in the cell give rise to disease phenotypes. Moreover, even with the unprecedented wealth of information available, obtaining such catalogues is extremely difficult.

Genetic Broadly speaking, genetic diseases are caused by aberrations in genes or chromosomes. Many genetic diseases are developed from before birth. Genetic disorders account for a significant number of the health care problems in our society. Advances in the understanding of this diseases have increased both the life span and quality of life for many of those affected by genetic disorders. Recent developments in bioinformatics and laboratory genetics have made possible the better delineation of certain malformation and mental retardation syndromes, so that their mode of inheritance can be understood. This information enables the genetic counselor to predict the risk for occurrence of a large number of genetic disorders. Most genetic counseling is done, however, only after the birth of at least one affected individual has alerted the family to their predilection for having children with a genetic disorder. The association of a single gene to a disease is rare and a genetic disease may or may not be a transmissible disorder. Some genetic diseases are inherited from the parent's genes, but others are caused by new mutations or changes to the DNA. In other occurrences, the same disease, for instance, some forms of carcinoma or melanoma, may stem from an inbred condition in some people, from new changes in other people, and from non-genetic causes in still other individuals.

There are more than six thousand known single-gene disorders (monogenic), which occur in about 1 out of every 200 births. As their term suggests, these diseases are caused by a mutation in one gene. By contrast, polygenic disorders are caused by several genes, regularly in combination with environmental factors. Examples of genetic phenotypes include Alzheimer's disease, breast cancer, leukemia, Down syndrome, heart defects, and deafness; therefore, cataloguing to sort out all the diseases related to genes is needed.

Challenges with Creation

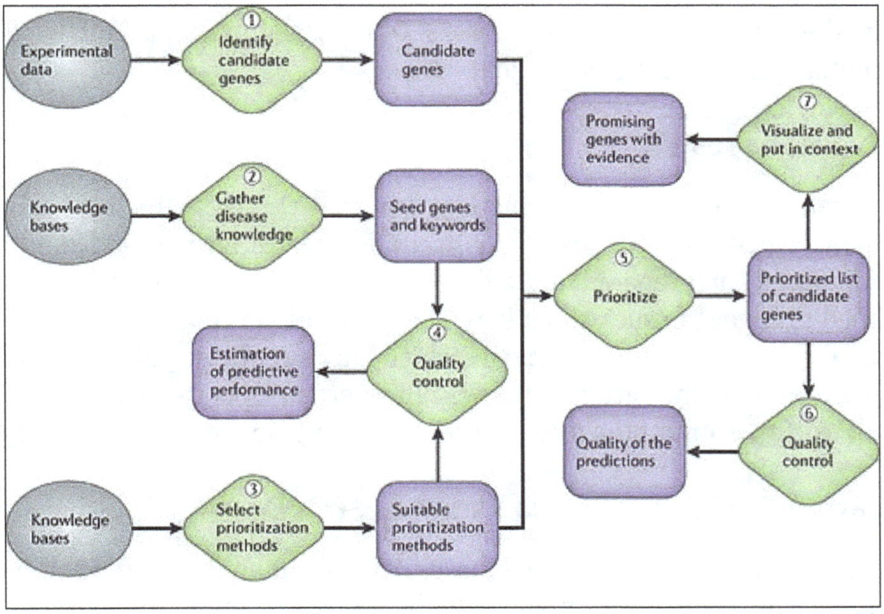

Gene prioritization workflow of human diseases: Typical lists come from linkage regions, chromosomal aberrations, association study loci, deferentially expressed gene lists or genes identified by sequencing variants. Alternatively, the complete genome can be prioritized, but substantially more false positives would then be expected.

At different stages of any gene disease project, molecular biologists need to choose, even after careful statistical data analysis, which genes or proteins to investigate further experimentally and which to leave out because of limited resources. Computational methods that integrate complex, heterogeneous data sets, such as expression data, sequence information, functional annotation and the biomedical literature, allow prioritizing genes for future study in a more informed way. Such methods can substantially increase the yield of downstream studies and are becoming invaluable to researchers. So one of the main concerns in biological and biomedical research is to recognise the underlying mechanisms behind this intricate genetic phenotypes. Great effort has been spent on finding the genes related to diseases.

However, increasingly evidences point out that most human diseases cannot be attributed to a single gene but arise due to complex interactions among multiple genetic variants and environmental risk factors. Several databases have been developed storing associations between genes and diseases such as the Comparative Toxicogenomics Database (CTD), Online Mendelian Inheritance in Man (OMIM), the genetic Association Database (GAD) or the Disease genetic Association Database (DisGeNET). Each of these databases focuses on different aspects of the phenotype-genotype relationship, and due to the nature of the database curation process, they are not complete, but in a way they are fully complementary between each other.

Types of Databases

Essentially, there are four types of databases: Curated databases, predictive databases, literature databases and integrative databases.

Curated Databases

The term curated data refers to information, that may comprise the most sophisticated computational formats for structured data, scientific updates, and curated knowledge, that has been composed and prepared under the regulation of one or more experts considered to be qualified to engage in such an activity The implication is that the resulting database is of high quality. The contrast is with data which may have been gathered through some automated process or using particularly low or inexpert unsupported data quality and possibly untrustworthy. Some of the most common examples include: CTD and UNIPROT.

Comparative Toxicogenomics Database (CTD)

The Comparative Toxicogenomics Database, helps to understand about the effects of environmental compounds on human health by integrating data from curated scientific literature to describe biochemical interactions with genes and proteins, and links between diseases and chemicals, and diseases and genes or proteins. CTD contains curated data defining cross-species chemical–gene/protein interactions and chemical– and gene–disease associations to illuminate molecular mechanisms underlying variable susceptibility and environmentally influenced diseases. These data deliver insights into complex chemical–gene and protein interaction networks. One of the main sources in this Database is curated information from OMIM.

CTD is a unique resource where bioinformatics specialists read the scientific literature and manually curate four types of core data:

- Chemical-gene interactions.
- Chemical-disease associations.
- Gene-disease associations.
- Chemical-phenotype associations.

Universal Protein Resource (UNIPROT)

The Universal Protein Resource (UniProt) is an inclusive resource for protein sequence and annotation data. It is a comprehensive, first-class and freely accessible database of protein sequence and functional information, that has many entries being derived from genome sequencing projects. It contains a large amount of information about the biological function of proteins derived from the study literature, which can hint to a direct connection between gene-protein-disease.

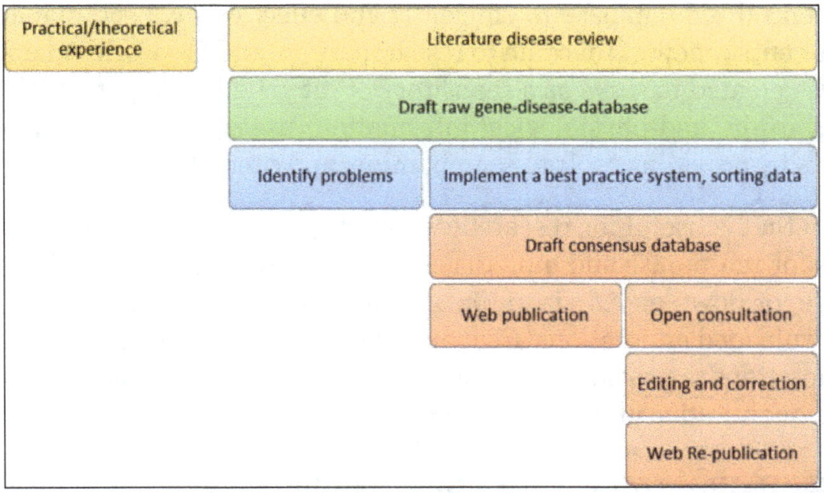

The process of database compilation and curation.The curated data may comprise a process from practical experience and literature review to web publication of the database.

Predictive Databases

A predictive database is one based on statistical inference. One particular approach to such inference is known as predictive inference, but the prediction can be undertaken within any of the several approaches to statistical inference. Indeed, one description of biostatistics is that it provides a means of transferring knowledge about a sample of a genetic population to the whole population (genomics), and to other related genes or genomes, which the same as prediction over time is not necessarily. When information is transferred across time, often to specific points in time, the process is known as forecasting. Three of the main examples of databases that can be considered in this category include: The Mouse genome Database (MGD), The Rat genome Database (RGD), OMIM and the SIFT Tool from Ensembl.

Mouse Genome Database (MGD)

The Mouse genome Database (MGD) is the international community resource for integrated genetic, genomic and biological data about the laboratory mouse. MGD provides full annotation of phenotypes and human disease associations for mouse models (genotypes) using terms from the Mammalian Phenotype Ontology and disease names from OMIM.

Rat Genome Database (RGD)

The Rat Genome Database (RGD) began as a collaborative effort between leading research institutions involved in rat genetic and genomic research. The rat continues to be extensively used by researchers as a model organism for investigating the biology and pathophysiology of disease. In the past several years, there has been a rapid increase in rat genetic and genomic data. This explosion of information highlighted the need for a centralized database to efficiently and effectively collect, manage, and distribute a rat-centric view of this data to researchers around the world. The Rat Genome Database was created to serve as a repository of rat genetic and genomic data, as well as mapping, strain, and physiological information. It also facilitates investigators research efforts by providing tools to search, mine, and predict this data.

Data at RGD that is useful for researchers investigating disease genes include disease annotations for rat, mouse and human genes. Annotations are manually curated from the literature, or downloaded via automated pipelines from other disease-related databases. Downloaded annotations are mapped to the same disease vocabulary used for manual annotations to provide consistency across the dataset. RGD also maintains disease-related quantitative phenotype data for the rat (PhenoMiner).

Online Mendelian Inheritance in Man (OMIM)

Supported by the NCBI, The Online Mendelian Inheritance in Man (OMIM) is a database that catalogues all the known diseases with a genetic component, and predicts their relationship to relevant genes in the human genome and provides references for further research and tools for genomic analysis of a catalogued gene. OMIM is a comprehensive, authoritative compendium of human genes and genetic phenotypes that is freely available and updated daily. The database has been used as a resource for predicting relevant information to inherited conditions.

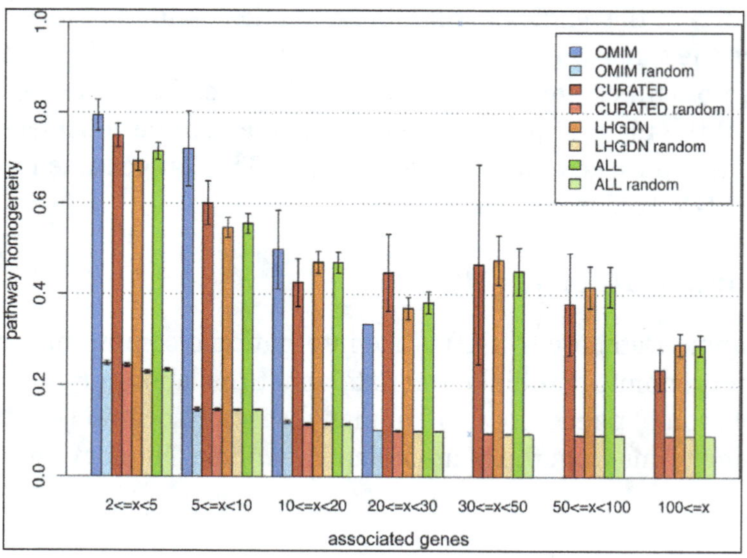

Pathway Hogeneity vs Associated Genes. Showing the concept that diseases have large association with a variety of genes, a mean pathway homogeneity values of single diseases and random controls are plotted for four networks binned by the number of associated gene products per disease. This graph shows how difficult is to correlate a bigger number of diseases vs concordance in 4 different databases, hence Gene Disease Databases test these relationships.

Ensembl SIFT Tool

This one of the largest resources available for all genomic and genetic studies, it provides a centralized resource for geneticists, molecular biologists and other researchers studying the genomes of our own species and other vertebrates and model disease organisms. Ensembl is one of several well-known genome browsers for the retrieval of genomic-disease information. Ensembl imports variation data from a variety of different sources, Ensembl predicts the effects of variants. For each variation that is mapped to the reference genome, each Ensembl transcript is identified that overlap the variation. Then it uses a rule-based approach to predict the effects that each allele of the variation may have on the transcript. The set of consequence terms, defined by the Sequence Ontology (SO) can be currently assigned to each combination of an allele and a transcript. Each allele of each variation may have a different effect in different transcripts. A variety of different tools are used to predict human mutations in the Ensembl database, one of the most widely used is SIFT, that predicts whether an amino acid substitution is likely to affect protein function based on sequence homology and the physic-chemical similarity between the alternate amino acids. The data provided for each amino acid substitution is a score and a qualitative prediction (either 'tolerated' or 'deleterious'). The score is the normalized probability that the amino acid change is tolerated so scores near 0 are more likely to be deleterious. The qualitative prediction is derived from this score such that substitutions with a score < 0.05 are called 'deleterious' and all others are called 'tolerated'. SIFT can be applied to naturally occurring nonsynonymous polymorphisms and laboratory-induced missense mutations, that will lead to build relationships in phenotype characteristics, proteomics and genomics.

Literature Databases

This sort of databases summarize books, articles, book reviews, dissertations, and annotations about gene-disease databases. Some of the following are examples of this type: GAD, LGHDN and BeFree Data.

Genetic Association Database (GAD)

The Genetic Association Database is an archive of human genetic association studies of complex diseases. GAD is primarily focused on archiving information on common complex human disease rather than rare Mendelian disorders as found in the OMIM. It includes curated summary data extracted from published papers in peer reviewed journals on candidate gene and genome Wide Association Studies (GWAS).

Literature-derived Human Gene-disease Network (LHGDN)

The literature-derived human gene-disease network (LHGDN) is a text mining derived database with focus on extracting and classifying gene-disease associations with respect to several biomolecular conditions. It uses a machine learning based algorithm to extract semantic gene-disease relations from a textual source of interest. It is part of the Linked Life Data, of the LMU in Munchen, Germany.

BeFree Data

Extracts gene-disease associations from MEDLINE abstract using the BeFree system. BeFree is composed of a biomedical Named Entity Recognition (BioNER) module to detect diseases and genes and a relation extraction module based on morphosyntactic information.

Integrative Databases

This sort of databases include Mendelian, compound and environmental diseases in an integrated gene-disease association archive and show that the concept of modularity applies for all of them They provide a functional analysis of diseases in case of important new biological insights, which might not be discovered when considering each of the gene-disease associations independently. Hence, they present a suitable framework for the study of how genetic and environmental factors, such as drugs, contribute to diseases. The best example for this sort of database is DisGeNET.

Gene Disease Associations Database DisGeNET

DisGeNET is a comprehensive gene-disease association database that integrates associations from several sources that covers different biomedical aspects of diseases. In particular, it is focused on the current knowledge of human genetic diseases including Mendelian, complex and environmental diseases. To assess the concept of modularity of human diseases, this database performs a systematic study of the emergent properties of human gene-disease networks by means of network topology and functional annotation analysis. The results indicate a highly shared genetic origin of human diseases and show that for most diseases, including Mendelian, complex and environmental diseases, functional modules exist. Moreover, a core set of biological pathways is found to be associated with most human diseases. Obtaining similar results when studying clusters of diseases, the findings in this database suggest that related diseases might arise due to dysfunction of common biological processes in the cell. The network analysis of this integrated database points out that data integration is needed to obtain a comprehensive view of the genetic landscape of human diseases and that the genetic origin of complex diseases is much more common than expected.

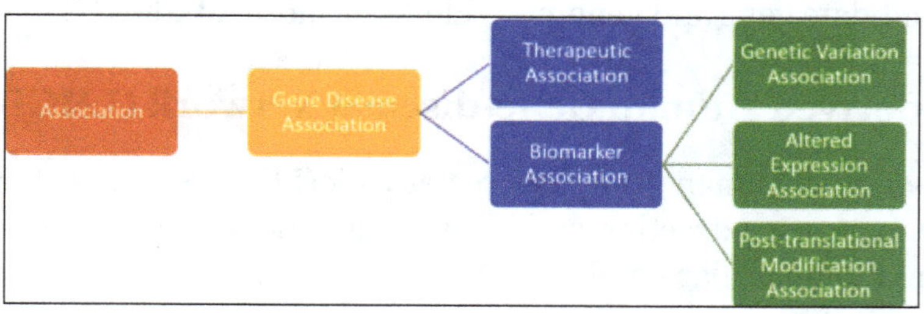

DisGeNET gene-disease association ontology. The description of each association type in this ontology is:

- Therapeutic Association: The gene/protein has a therapeutic role in the amelioration of the disease. #Biomarker Association: The gene/protein either plays a role in the etiology of the disease (e.g. participates in the molecular mechanism that leads to disease) or is a biomarker for a disease.
- Genetic Variation Association: Used when a sequence variation (a mutation, a SNP) is associated to the disease phenotype, but there is still no evidence to say that the variation causes the disease. In some cases the presence of the variants increase the susceptibility to the disease. In general, the NCBI SNP identifiers are provided.
- Altered Expression Association: Alterations in the function of the protein by means of altered expression of the gene are associated with the disease phenotype.
- Post-translational Modification Association: Alterations in the function of the protein by means of post-translational modifications (methylation or phosphorylation of the protein) are associated with the disease phenotype.

GENE EXPRESSION OMNIBUS DATABASE

The Gene Expression Omnibus (GEO) is an online database for microarray, next generation sequencing and other forms of high–throughput functional genomic data. GEO also contains different options to retrieve and study the various gene expression data submitted by different scientist. The data needs to be submitted to MIAME (Minimum Information about a Microarray Experiment), complaint public repository like GEO, where they provide assistance so as to present the data for journal publication. Minimum Information About a Microarray Experiment depicts the minimum information for the microarray data which can be easily interpreted and analyzed. It is formed by FGED Society for reporting microarray experiments. Microarray is a collection of microscopic DNA sequence spots attached to a solid surface of glass or silicon. Each spot contains millions of replicates of specific DNA sequences called probes, which are short section of a gene or other DNA element. The microarray technology is a combination of wet-lab techniques, statistical analysis and application of informatics to the data. The data obtained is submitted to GEO. The 3 main objectives of GEO are Submission guide, Data organization and Query and Analysis.

Data Organization

The data from expression studies can be organised into platform records, sample records, series records, dataset and profiles.

Platform records are submitted by the scientist for curation, which contains many samples from multiple submitters. It contains information about the array/sequencer. For array based platforms, a data table with its array template are provided. Each of that record is assigned with a unique and stable GEO accession number (GPLxxx).

Sample record is assigned with a unique and stable GEO accession number (GSMxxx), which contains the sample, the changes and manipulation it has undergone and the abundance measurement of each element derived from it.

Series record is assigned with a unique and stable GEO accession number (GSExxx), which is an original record from the submitter summarizing an experiment. It contains a group of related samples with its description, tables, extracted data, summary, conclusion and analysis. The data submitted by the scientist are examined by various curators in GEO before going for a paper publication or any analysis.

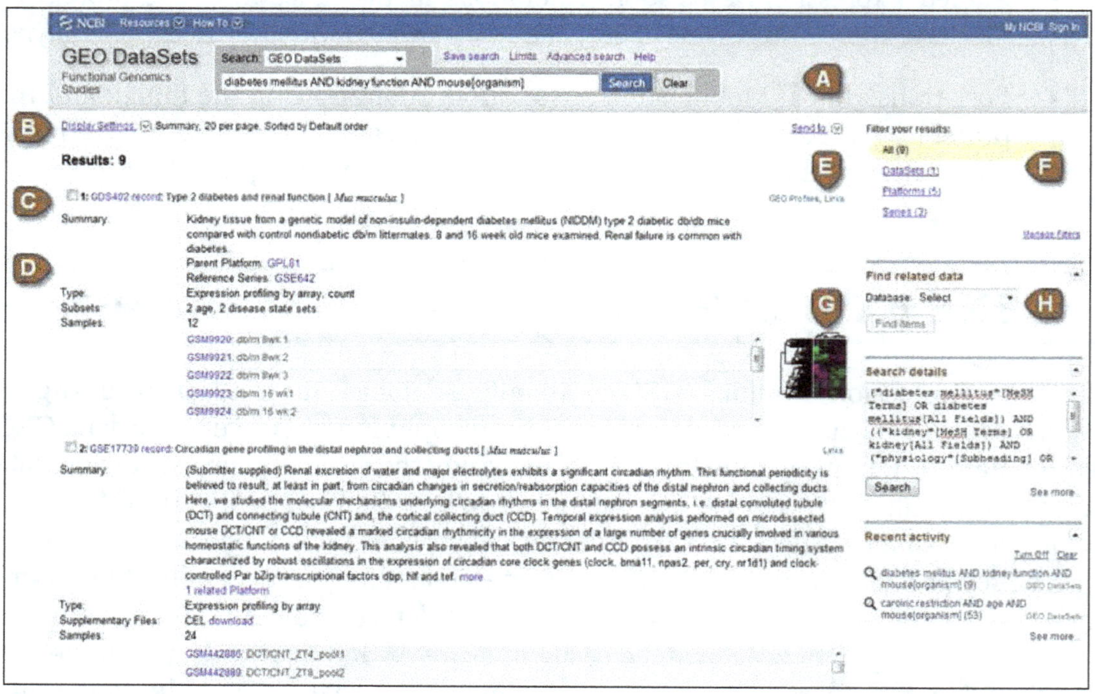

Dataset.

GEO Dataset: The data which are selected after curation are refined again before transferred into GEO Dataset and Gene profile records. GEO Dataset is a study based database which contains Platform, Samples, Series and the corrected data from the curators. These curated dataset form the root of GEO's advanced data display and analysis features, with gene expression level identifying tools and cluster heat maps. These dataset are assigned with a unique and stable GEO accession number GDSxxx. It can be searched by simply typing any keywords or Organism name, authors or Dataset type in their respective boxes. A large amount of data can be filtered down to make our search more specific by using GEO Limits Query builder and Advanced Search. GEO

Dataset contains GEO samples which refer to the same platform with same set of array elements. The values are calculated by background processing and normalization which are consistent across the dataset subsets. Both GEO dataset and GEO series can be searched using the same dataset interface. It is not possible to represent all the submitted data in a dataset because every day the curators are experiencing a backlog in dataset creation.

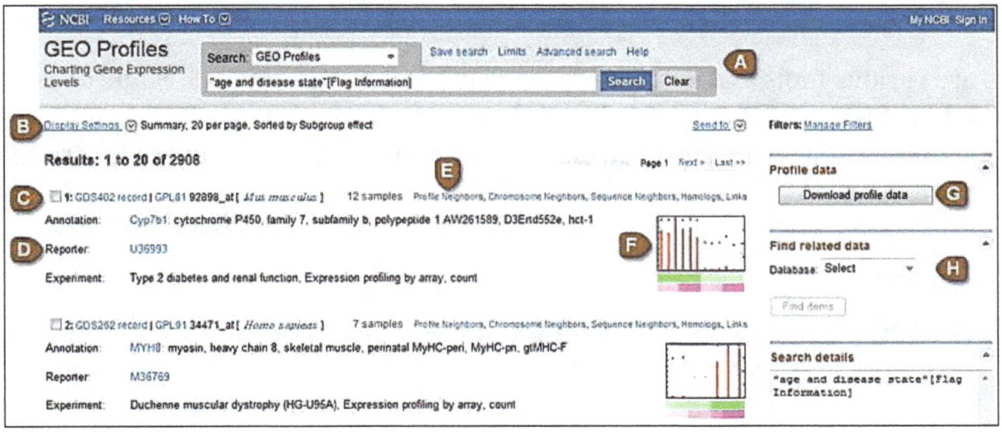

Example for profile.

GEO Profiles are derived from curated GEO datasets. GEO Profile is a gene based database where the user can search for gene expression profiles. Every single profile is represented as a chart which is displaying the expression level of one gene across all samples within a Dataset. Experimental details are represented as bars along the bottom of the charts. It is also possible to view the similarly expressed genes within close proximity on the chromosome. All the profiles are connected with internal links to represent similarly behavioring genes and external links to significant records in other NCBI databases. It can be searched with any keywords, gene names, gene symbols, GenBank accession number or profiles flagged as being differently expressed.

References

- Curry E, Freitas A, O'Riáin S (2010). "The Role of Community-Driven Data Curation for Enterprises". In D. Wood (ed.). Linking Enterprise Data. Boston, MA: Springer US. pp. 25–47. ISBN 978-1-441-97664-2
- Biological-databases-types-and-importance: microbenotes.com, Retrieved 16 June, 2020
- Sikic, K.; Carugo, O. (2010). "Protein sequence redundancy reduction: comparison of various method". Bioinformation. 5 (6): 234–9. doi:10.6026/97320630005234. PMC 3055704. PMID 21364823
- Ensemble-clustering-for-biological-datasets, bioinformatics, books: intechopen.com, Retrieved 13 April, 2020
- "Hemoglobin beta locus or HBB (141900): Allelic variants". Online Mendelian Inheritance in Man. Baltimore, MD: McKusick-Nathans Institute of Genetic Medicine, Johns Hopkins University. 2015. Retrieved 23 July 2015

Bioinformatics Tools and Software

5 CHAPTER

There are various tools and software that are used in bioinformatics. Some of them are Clustal, PHYLIP, FASTA, BLAST, BLAT, BioSLAX, BioPerl, BioJava, ArrayTrack, etc. This chapter closely examines the different bioinformatics tools and software to provide an extensive understanding of the subject.

BIOINFORMATICS TOOLS

Molecular Biology

- CRISPR guide RNA design tools.
- gRNA design tool.
- Genome-wide gRNA Database.
- Gene Mutagenesis designer.
- Restriction Enzyme Map Analysis.
- Restriction Enzyme Tool.
- siRNA Target Finder.
- siRNA construct Builder.
- siRNA Sequence Scrambler.
- GenSmart Codon Optimization Tool.
- Rare Codon Analysis Tool.
- Codon Frequency Tables.
- Genetic Code – Codon Tables.
- The Sequence Manipulation Suite.
- GenPool ORF & Full-length Gene Database.

- Real-time PCR (TaqMan) Primer Design.
- Oligo Calculator for Tm, MV, and ug/OD.
- Primer Design.
- Sequencing Primer Design.
- DNA Construct Design Tool.
- GenSmart Design.

Peptide

- Chart of Amino Acid Structure.
- Peptide Molecular Weight Calculator.
- Amino Acid Code.
- Peptide Property Calculator.
- Peptide Screening Tools.
- Antigen Prediction.

Common Tools

- Nucleic Acids Conversions.
- Molecular Biology Glossary.
- Antibody Drug Discovery Glossary.

Protein

- Psort II.
- WOLF-PSORT.

OPEN READING FRAME

In molecular genetics, an open reading frame (ORF) is the part of a reading frame that has the ability to be translated. An ORF is a continuous stretch of codons that begins with a start codon (usually AUG) and ends at a stop codon (usually UAA, UAG or UGA). An ATG codon (AUG in terms of RNA) within the ORF (not necessarily the first) may indicate where translation starts. The transcription termination site is located after the ORF, beyond the translation stop codon. If transcription were to cease before the stop codon, an incomplete protein would be made during translation. In eukaryotic genes with multiple exons, introns are removed and exons are then joined together after transcription to yield the final mRNA for protein translation. In the context of gene finding, the start-stop definition of an ORF therefore only applies to spliced mRNAs, not genomic DNA, since introns may contain stop codons and cause shifts between reading frames. An alternative definition says that an ORF is a sequence that has a length divisible by three and is bounded by stop codons. This more general definition can also be useful in the context of transcriptomics and metagenomics, where start and stop codon may not be present in the obtained sequences. Such an ORF corresponds to parts of a gene rather than the complete gene.

```
1. ATG CAA TGG GGA AAT GTT ACC AGG TCC GAA CTT ATT GAG GTA AGA CAG ATT TAA
2. A TGC AAT GGG GAA ATG TTA CCA GGT CCG AAC TTA TTG AGG TAA GAC AGA TTT AA
3. AT GCA ATG GGG AAA TGT TAC CAG GTC CGA ACT TAT TGA GGT AAG ACA GAT TTA A
```

Sample sequence showing three different possible reading frames. Start codons are highlighted in purple, and stop codons are highlighted in red.

Biological Significance

One common use of open reading frames (ORFs) is as one piece of evidence to assist in gene prediction. Long ORFs are often used, along with other evidence, to initially identify candidate protein-coding regions or functional RNA-coding regions in a DNA sequence. The presence of an ORF does not necessarily mean that the region is always translated. For example, in a randomly generated DNA sequence with an equal percentage of each nucleotide, a stop-codon would be expected once every 21 codons. A simple gene prediction algorithm for prokaryotes might look for a start codon followed by an open reading frame that is long enough to encode a typical protein, where the codon usage of that region matches the frequency characteristic for the given organism's coding regions. Therefore, some authors say that an ORF should have a minimal length, e.g. 100 codons or 150 codons. By itself even a long open reading frame is not conclusive evidence for the presence of a gene. On the other hand, it has been proven that some short ORFs (sORFs) that lack the classical hallmarks of protein-coding genes (both from ncRNAs and mRNAs) can produce functional peptides. 5'NTR of about 50% of mammal mRNAs are known to contain one or several sORFs. 64–75% of experimentally found translation initiation sites of sORFs are conserved in the genomes of human and mouse and may indicate that these elements have function. However, sORFs can often be found only in the minor forms of mRNAs and avoid the selection; the high conservatism of initiation sites may be connected with their location inside promoters of the relevant genes. Such kind of situation is characteristic of SLAMF1 gene, for example.

Six-frame Translation

Since DNA is interpreted in groups of three nucleotides (codons), a DNA strand has three distinct reading frames. The double helix of a DNA molecule has two anti-parallel strands; with the two strands having three reading frames each, there are six possible frame translations.

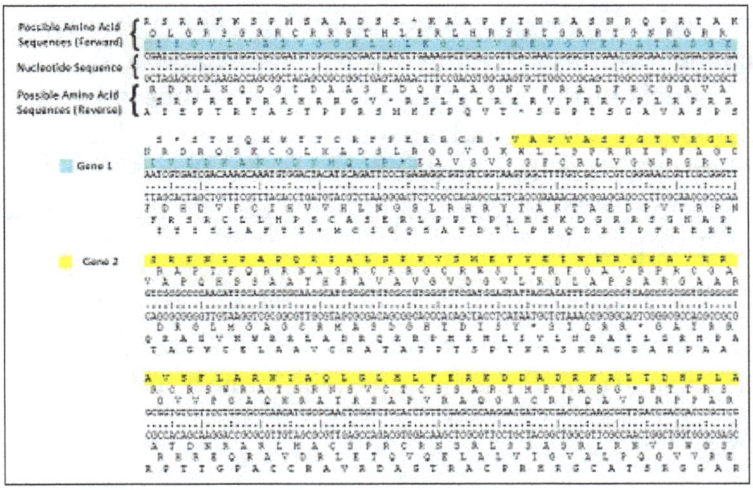

Example of a six-frame translation. The nucleotide sequence is shown in the middle with forward translations above and reverse translations below. Two possible open reading frames with the sequences are highlighted.

ORF Finding Tools

ORF Finder

The ORF Finder (Open Reading Frame Finder) is a graphical analysis tool which finds all open reading frames of a selectable minimum size in a user's sequence or in a sequence already in the database. This tool identifies all open reading frames using the standard or alternative genetic codes. The deduced amino acid sequence can be saved in various formats and searched against the sequence database using the BLAST server. The ORF Finder should be helpful in preparing complete and accurate sequence submissions. It is also packaged with the Sequin sequence submission software (Sequence analyser).

ORF Investigator

ORF Investigator is a program which not only gives information about the coding and non coding sequences but also can perform pairwise global alignment of different gene/DNA regions sequences. The tool efficiently finds the ORFs for corresponding amino acid sequences and converts them into their single letter amino acid code, and provides their locations in the sequence. The pairwise global alignment between the sequences makes it convenient to detect the different mutations, including single nucleotide polymorphism. Needleman–Wunsch algorithms are used for the gene alignment. The ORF Investigator is written in the portable Perl programming language, and is therefore available to users of all common operating systems.

ORF Predictor

OrfPredictor is a web server designed for identifying protein-coding regions in expressed sequence tag (EST)-derived sequences. For query sequences with a hit in BLASTX, the program predicts the coding regions based on the translation reading frames identified in BLASTX alignments, otherwise, it predicts the most probable coding region based on the intrinsic signals of the query sequences. The output is the predicted peptide sequences in the FASTA format, and a definition line that includes the query ID, the translation reading frame and the nucleotide positions where the coding region begins and ends. OrfPredictor facilitates the annotation of EST-derived sequences, particularly, for large-scale EST projects.

ORF Predictor uses a combination of the two different ORF definitions mentioned above. It searches stretches starting with a start codon and ending at a stop codon. As an additional criterion, it searches for a stop codon in the 5' untranslated region (UTR).

ORFik

ORFik is a R-package in Bioconductor for finding open reading frames and using Next generation sequencing technologies for justification of ORFs.

CLUSTAL

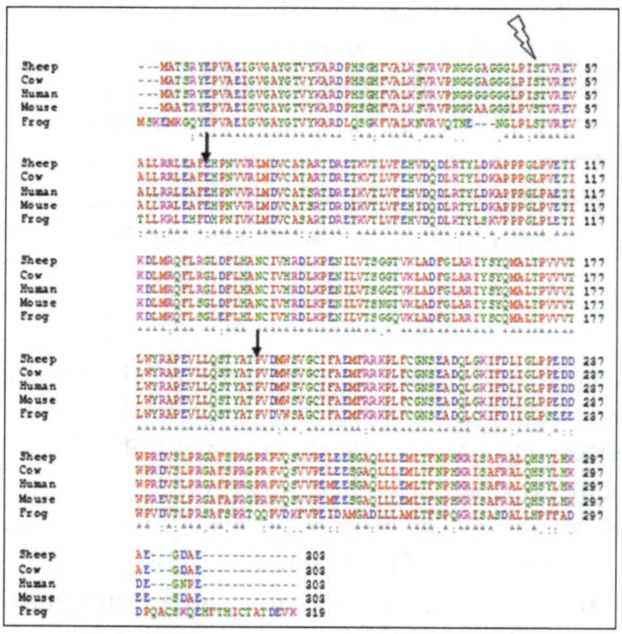

Multiple sequence alignment of CDK4 protein generated with ClustalW. Arrows indicate point mutations.

Clustal is a series of widely used computer programs used in Bioinformatics for multiple sequence alignment. There have been many versions of Clustal over the development of the algorithm that are listed below. The analysis of each tool and its algorithm are also detailed in their respective categories. Available operating systems listed in the sidebar are a combination of the software availability and may not be supported for every current version of the Clustal tools. Clustal Omega has the widest variety of operating systems out of all the Clustal tools.

There have been many variations of the Clustal software, all of which are listed below:

- Clustal: The original software for multiple sequence alignments, created by Des Higgins in 1988, was based on deriving phylogenetic trees from pairwise sequences of amino acids or nucleotides.
- ClustalV: The second generation of the Clustal software was released in 1992 and was a rewrite of the original Clustal package. It introduced phylogenetic tree reconstruction on the final alignment, the ability to create alignments from existing alignments, and the option to create trees from alignments using a method called Neighbor joining.
- ClustalW: The third generation, released in 1994, greatly improved upon the previous versions. It improved upon the progressive alignment algorithm in various ways, including allowing individual sequences to be weighted down or up according to similarity or divergence respectively in a partial alignment. It also included the ability to run the program in batch mode from the command line.
- ClustalX: This version, released in 1997, was the first to have a graphical user interface.
- ClustalΩ (Omega): The current standard version.
- Clustal2: The updated versions of both ClustalW and ClustalX with higher accuracy and efficiency.

Name Origin

The guide tree in the initial programs was constructed via a UPGMA cluster analysis of the pairwise alignments, hence the name CLUSTAL. The first four versions in 1988 had Arabic numerals (1 to 4), whereas with the fifth version Des Higgins switched to Roman numeral V in 1992. In 1994 and in 1997, for the next two versions, the letters after the letter V were used and made to correspond to W for Weighted and X for X Window. The name omega was chosen to mark a change from the previous ones.

Function

All variations of the Clustal software align sequences using a heuristic that progressively builds a multiple sequence alignment from a series of pairwise alignments. This method works by analyzing the sequences as a whole, then utilizing the UPGMA/Neighbor-joining method to generate a distance matrix. A guide tree is then calculated from the scores of the sequences in the matrix, then subsequently used to build the multiple sequence alignment by progressively aligning the sequences in order of similarity. Essentially, Clustal creates multiple sequence alignments through three main steps:

- Do a pairwise alignment using the progressive alignment method.
- Create a guide tree (or use a user-defined tree).
- Use the guide tree to carry out a multiple alignment.

These steps are carried out automatically when you select "Do Complete Alignment". Other options are "Do Alignment from guide tree and phylogeny" and "Produce guide tree only".

Input/Output

This program accepts a wide range of input formats, including NBRF/PIR, FASTA, EMBL/Swiss-Prot, Clustal, GCC/MSF, GCG9 RSF, and GDE. The output format can be one or many of the following: Clustal, NBRF/PIR, GCG/MSF, PHYLIP, GDE, or NEXUS.

Reading Multiple Sequence Alignment Output		
Symbol	Definition	Meaning
*	asterisk	positions that have a single and fully conserved residue
:	colon	conservation between groups of strongly similar properties with a score greater than.5 on the PAM 250 matrix
.	period	conservation between groups of weakly similar properties with a score less than or equal to.5 on the PAM 250 matrix

The same symbols are shown for both DNA/RNA alignments and protein alignments, so while * (asterisk) symbols are useful to both, the other consensus symbols should be ignored for DNA/RNA alignments.

Settings

Many settings can be modified to adapt the alignment algorithm to different circumstances. The main parameters are the gap opening penalty, and the gap extension penalty.

Clustal and ClustalV

The original program in the Clustal series of software was developed in 1988 as a way to generate multiple sequence alignments on personal computers. ClustalV was released 4 years later and greatly improved upon the original, adding and altering a few key features, including a switch to being written in C instead of Fortran like its predecessor.

Algorithm

Both versions use the same fast approximate algorithm to calculate the similarity scores between sequences, which in turn produces the pairwise alignments. The algorithm works by calculating the similarity scores as the number of k-tuple matches between two sequences, accounting for a set penalty for gaps. The more similar the sequences, the higher the score, the more divergent, the lower the scores. Once the sequences are scored, a dendrogram is generated through the UPGMA to represent the ordering of the multiple sequence alignment. The higher ordered sets of sequences are aligned first, followed by the rest in descending order. The algorithm allows for very large data sets, and works fast. However, the speed is dependent on the range for the k-tuple matches chosen for the particular sequence type.

Notable ClustalV Improvements

Some of the most notable additions in ClustalV are profile alignments, and full command line interface options. The ability to use profile alignments allows the user to align two or more previous alignments or sequences to a new alignment and move misaligned sequences (low scored) further down the alignment order. This gives the user the option to gradually and methodically create multiple sequence alignments with more control than the basic option. The option to run from the command line greatly expedites the multiple sequence alignment process. Sequences can be run with a simple command,

```
clustalv nameoffile.seq
```

or,

```
clustalv /infile=nameoffile.seq
```

and the program will determine what type of sequence it is analyzing. When the program is completed, the output of the multiple sequence alignment as well as the dendrogram go to files with.aln and.dnd extensions respectively. The command line interface uses the default parameters, and doesn't allow for other options.

ClustalW

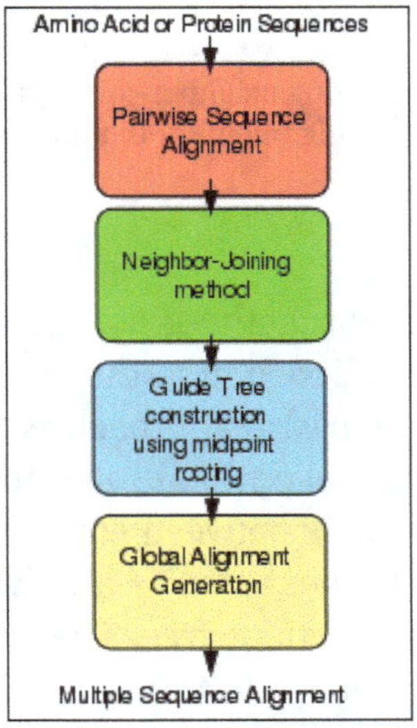

Depicts the steps the ClustalW software algorithm uses for global alignments.

ClustalW like the other Clustal tools is used for aligning multiple nucleotide or protein sequences in an efficient manner. It uses progressive alignment methods, which align the most similar sequences first and work their way down to the least similar sequences until a global alignment is created. ClustalW is a matrix-based algorithm, whereas tools like T-Coffee and Dialign are consistency-based. ClustalW has a fairly efficient algorithm that competes well against other software. This program requires three or more sequences in order to calculate a global alignment, for pairwise sequence alignment (2 sequences) use tools similar to EMBOSS, LALIGN.

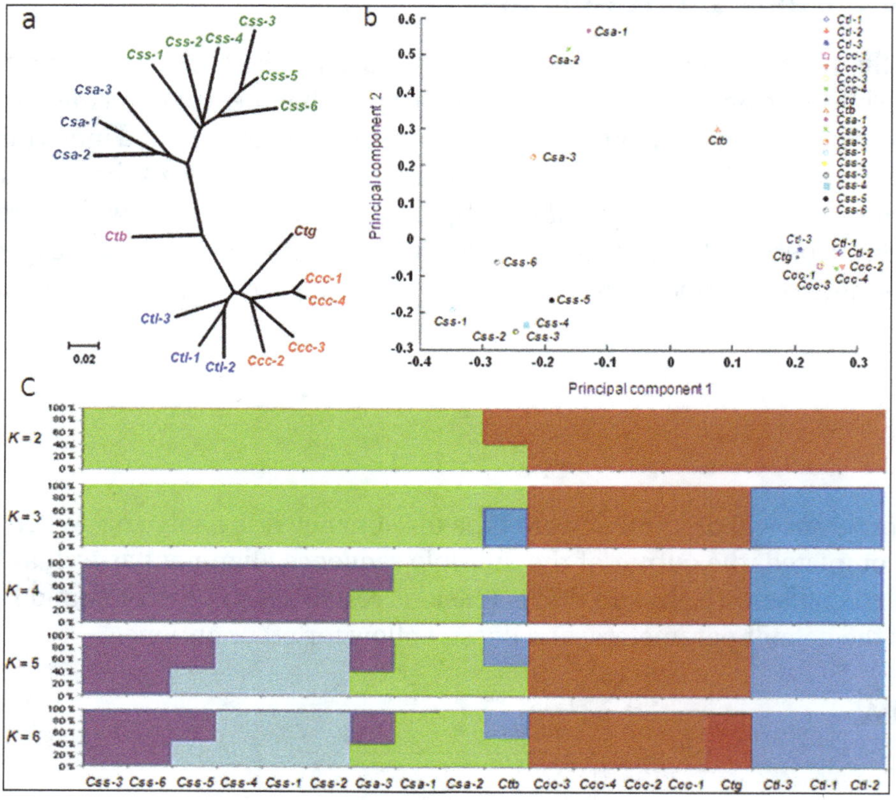

Diagram showing neighbor-joining method in sequence alignment for bioinformatics.

Algorithm

ClustalW uses progressive alignment methods as stated above. In these, the sequences with the best alignment score are aligned first, then progressively more distant groups of sequences are aligned. This heuristic approach is necessary due to the time and memory demand of finding the global optimal solution. The first step to the algorithm is computing a rough distance matrix between each pair of sequences, also known as pairwise sequence alignment. The next step is a neighbor-joining method that uses midpoint rooting to create an overall guide tree. The process it uses to do this is shown in the detailed diagram for the method to the right. The guide tree is then used as a rough template to generate a global alignment.

Time Complexity

ClustalW has a time complexity of $O(N^2)$ because of its use of the neighbor-joining method. In the updated version (ClustalW2) there is an option built into the software to use UPGMA which is faster with large input sizes. The command line flag in order to use it instead of neighbor-joining is:

```
-clustering=UPGMA
```

For example, on a standard desktop, running UPGMA on 10,000 sequences would produce results in less than a minute while neighbor-joining would take over an hour. By running the ClustalW algorithm with this adjustment, it saves significant amounts of time. ClustalW2 also has an option to use iterative alignment to increase alignment accuracy. While it is not necessarily faster or more efficient complexity-wise, the increase in accuracy is valuable and can be useful for smaller data sizes. These are the various command line flags to achieve this:

```
-Iteration=Alignment

-Iteration=Tree

-numiters
```

The first command line option refines the final alignment. The second option incorporates the scheme into the progressive alignment step of the algorithm. The third specifies the number of iteration cycles where the default value is set to 3.

Accuracy and Results

The algorithm ClustalW uses provides a close-to-optimal result almost every time. However, it does exceptionally well when the data set contains sequences with varied degrees of divergence. This is because in a data set like this, the guide tree becomes less sensitive to noise. ClustalW was one of the first algorithms to combine pairwise alignment and global alignment in an attempt to be speed efficient, and it worked, but there is a loss in accuracy that other software doesn't have due to this.

ClustalW, when compared to other MSA algorithms, performed as one of the quickest while still maintaining a level of accuracy. There is still much to be improved compared to its consistency-based competitors like T-Coffee. The accuracy for ClustalW when tested against MAFFT, T-Coffee, Clustal Omega, and other MSA implementations had the lowest accuracy for full-length sequences. It had the least RAM memory demanding algorithm out of all the ones tested in the study. While ClustalW recorded the lowest level of accuracy among its competitors, it still maintained what some would deem acceptable. There have been updates and improvements to the algorithm that are present in ClustalW2 that work to increase accuracy while still maintaining its greatly valued speed.

Clustal Omega

ClustalΩ (alternatively written as Clustal O and Clustal Omega) is a fast and scalable program written in C and C++ used for multiple sequence alignment. It uses seeded guide trees and a new HMM engine that focuses on two profiles to generate these alignments. The program requires three or more sequences in order to calculate the multiple sequence alignment, for two sequences use pairwise sequence alignment tools (EMBOSS, LALIGN). Clustal Omega is consistency-based and is widely viewed as one of the fastest online implementations of all multiple sequence alignment tools and still ranks high in accuracy, among both consistency-based and matrix-based algorithms.

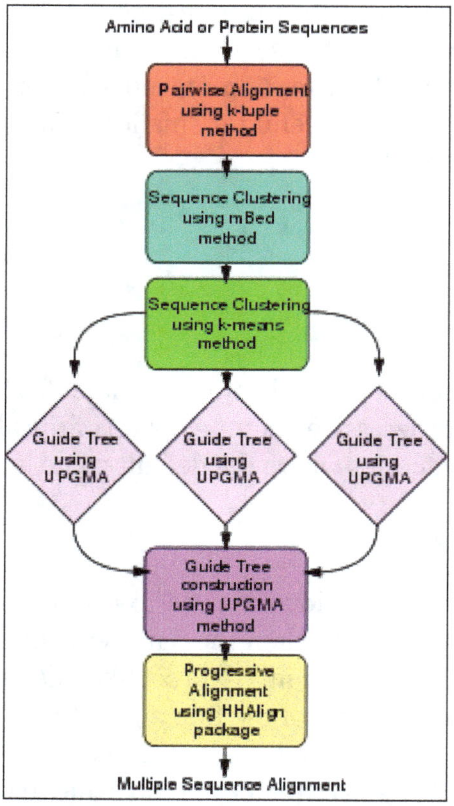

Flowchart depicting the step-by-step algorithm used in Clustal Omega.

Algorithm

Clustal Omega has five main steps in order to generate the multiple sequence alignment. The first is producing a pairwise alignment using the k-tuple method, also known as the word method. This, in summary, is a heuristic method that isn't guaranteed to find an optimal alignment solution, but is significantly more efficient than the dynamic programming method of alignment. After that, the sequences are clustered using the modified mBed method. The mBed method calculates pairwise distance using sequence embedding. This step is followed by the k-means clustering method. Next, the guide tree is constructed using the UPGMA method. This is shown as multiple guide tree steps leading into one final guide tree construction because of the way the UPGMA algorithm works. At each step, (each diamond in the flowchart) the nearest two clusters are combined and is repeated until the final tree can be assessed. In the final step, the multiple sequence alignment is produced using HHAlign package from the HH-Suite, which uses two profile HMM's. A profile HMM is a linear state machine consisting of a series of nodes, each of which corresponds roughly to a position (column) in the alignment from which it was built.

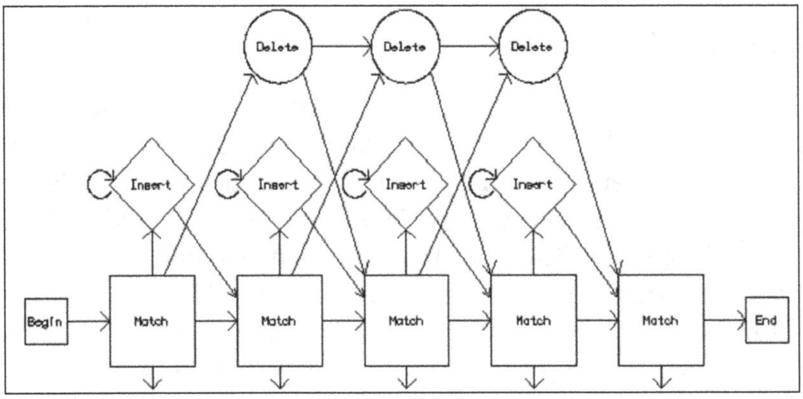

The structure of a profile HMM used in the implementation of Clustal Omega.

Time Complexity

The exact way of computing an optimal alignment between N sequences has a computational complexity of $O(L^N)$ for N sequences of length L making it prohibitive for even small numbers of sequences. Clustal Omega uses a modified version of mBed which has a complexity of $O(N \log N)$, and produces guide trees that are just as accurate as those from conventional methods. The speed and accuracy of the guide trees in Clustal Omega is attributed to the implementation of a modified mBed algorithm. It also reduces the computational time and memory requirements to complete alignments on large datasets.

Accuracy and Results

The accuracy of Clustal Omega on a small number of sequences is, on average, very similar to what are considered high quality sequence aligners. The difference comes when using large sets of data with hundreds of thousands of sequences. In these cases, Clustal Omega outperforms other algorithms across the board. Its completion time and overall quality is consistently better than other programs. It is capable of running 100,000+ sequences on one processor in a few hours.

Clustal Omega uses the HHAlign package of the HH-Suite, which aligns two profile Hidden Markov Models instead of a profile-profile comparison. This improves the quality of the sensitivity and alignment significantly. This, combined with the mBed method, gives Clustal Omega its advantage over other sequence aligners. The results end up being very accurate and very quick which is the optimal situation.

On data sets with nonconserved terminal bases, Clustal Omega may be more accurate than Probcons and T-Coffee despite the fact that both of these are consistency-based algorithms, in contrast to Clustal Omega. On an efficiency test with programs that produce high accuracy scores, MAFFT was the fastest, closely followed by Clustal Omega. Both were faster than T-Coffee, however, MAFFT and Clustal Omega required more memory to run.

Clustal2 (ClustalW/ClustalX)

Clustal2 is the packaged release of both the command-line ClustalW and graphical Clustal X. Neither are new tools, but are updated and improved versions of the previous implementations seen above. Both downloads come precompiled for many operating systems like Linux, Mac OS X and Windows (both XP and Vista). This release was designed in order to make the website more organized and user friendly, as well as updating the source codes to their most recent versions. Clustal2 is version 2 of both ClustalW and ClustalX, which is where it gets its name. Past versions can still be found on the website, however, every precompilation is now up to date.

PHYLIP

PHYLogeny Inference Package (PHYLIP) is a free computational phylogenetics package of programs for inferring evolutionary trees (phylogenies). It consists of 35 portable programs, i.e., the source code is written in the programming language C. As of version 3.696, it is licensed as open-source software; versions 3.695 and older were proprietary software freeware. Releases occur as source code, and as precompiled executables for many operating systems including Windows (95, 98, ME, NT, 2000, XP, Vista), Mac OS 8, Mac OS 9, OS X, Linux (Debian, Red Hat); and FreeBSD from FreeBSD.org. Full documentation is written for all the programs in the package and is included therein. The author of the package is Professor Joseph Felsenstein, of the Department of Genome Sciences and the Department of Biology, University of Washington, Seattle.

Methods (implemented by each program) that are available in the package include parsimony, distance matrix, and likelihood methods, including bootstrapping and consensus trees. Data types that can be handled include molecular sequences, gene frequencies, restriction sites and fragments, distance matrices, and discrete characters.

Each program is controlled through a menu, which asks users which options they want to set, and allows them to start the computation. The data is read into the program from a text file, which the user can prepare using any word processor or text editor (but this text file cannot be in the special format of that word processor, it must instead be in flat ASCII or text only format). Some sequence analysis programs such as the ClustalW alignment program can write data files in the PHYLIP format. Most of the programs look for the data in a file called infile. If they do not find this file, they then ask the user to type in the file name of the data file.

Output is written onto files with names like outfile and outtree. Trees written onto outtree are in the Newick format, an informal standard agreed to in 1986 by authors of seven major phylogeny packages.

FASTA

FASTA is a DNA and protein sequence alignment software package first described by David J. Lipman and William R. Pearson in 1985. Its legacy is the FASTA format which is now ubiquitous in bioinformatics.

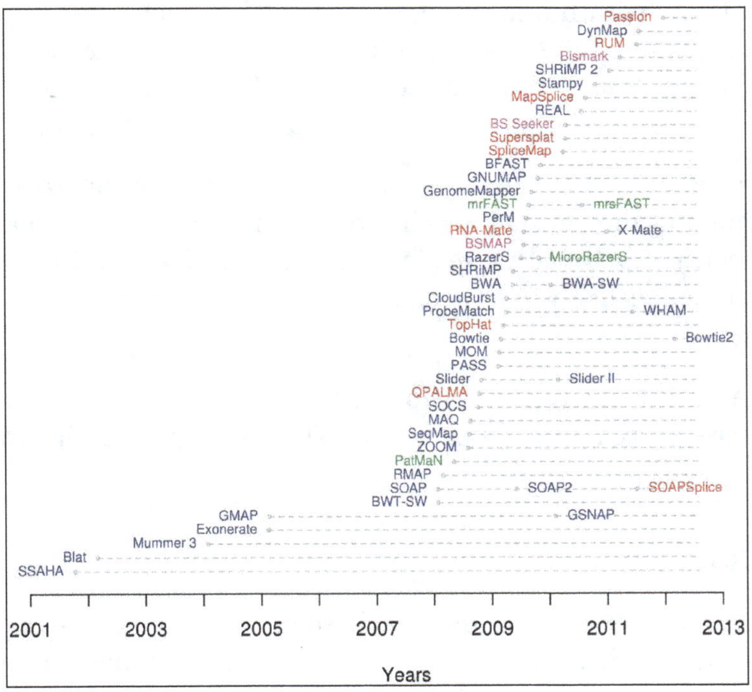

Mappers timeline. DNA mappers are plotted in blue, RNA mappers in red, miRNA mappers in green and bisulphite mappers in purple. Grey dotted lines connect related mappers (extensions or new versions). The timeline only includes mappers with peer-reviewed publications, and the date corresponds to the earliest date of publication (e.g. advanced publication date as opposed to the date of publication).

The original FASTP program was designed for protein sequence similarity searching. Because of the exponentially expanding genetic information and the limited speed and memory of computers in the 1980s heuristic methods were introduced aligning a query sequence to entire data-bases. FASTA, published in 1987, added the ability to do DNA:DNA searches, translated protein:DNA searches, and also provided a more sophisticated shuffling program for evaluating statistical significance. There are several programs in this package that allow the alignment of protein sequences and DNA sequences. Nowadays, increased computer performance makes it possible to perform searches for local alignment detection in a database using the Smith–Waterman algorithm.

FASTA is pronounced "fast A", and stands for "FAST-All", because it works with any alphabet, an extension of the original "FAST-P" (protein) and "FAST-N" (nucleotide) alignment tools.

Uses

The current FASTA package contains programs for protein:protein, DNA:DNA, protein:translated DNA (with frameshifts), and ordered or unordered peptide searches. Recent versions of the FASTA package include special translated search algorithms that correctly handle frameshift errors (which six-frame-translated searches do not handle very well) when comparing nucleotide to protein sequence data.

In addition to rapid heuristic search methods, the FASTA package provides SSEARCH, an implementation of the optimal Smith–Waterman algorithm. A major focus of the package is the calculation of accurate similarity statistics, so that biologists can judge whether an alignment is likely to have occurred by chance, or whether it can be used to infer homology.

The FASTA file format used as input for this software is now largely used by other sequence database search tools (such as BLAST) and sequence alignment programs (Clustal, T-Coffee, etc.).

Search Method

FASTA takes a given nucleotide or amino acid sequence and searches a corresponding sequence database by using local sequence alignment to find matches of similar database sequences.

The FASTA program follows a largely heuristic method which contributes to the high speed of its execution. It initially observes the pattern of word hits, word-to-word matches of a given length, and marks potential matches before performing a more time-consuming optimized search using a Smith–Waterman type of algorithm.

The size taken for a word, given by the parameter kmer, controls the sensitivity and speed of the program. Increasing the kmer value decreases number of background hits that are found. From the word hits that are returned the program looks for segments that contain a cluster of nearby hits. It then investigates these segments for a possible match.

There are some differences between fastn and fastp relating to the type of sequences used but both use four steps and calculate three scores to describe and format the sequence similarity results. These are:

- Identify regions of highest density in each sequence comparison. Taking a kmer to equal 1 or 2.

In this step all or a group of the identities between two sequences are found using a look up table. The kmer value determines how many consecutive identities are required for a match to be declared. Thus the lesser the kmer value: the more sensitive the search. kmer=2 is frequently taken by users for protein sequences and kmer=4 or 6 for nucleotide sequences. Short oligonucleotides are usually run with kmer= 1. The program then finds all similar local regions, represented as diagonals of a certain length in a dot plot, between the two sequences by counting kmer matches and penalizing for intervening mismatches. This way, local regions of highest density matches in a diagonal are isolated from background hits. For protein sequences BLOSUM50 values are used for scoring kmer matches. This ensures that groups of identities with high similarity scores contribute more to the local diagonal score than to identities with low similarity scores. Nucleotide sequences use the identity matrix for the same purpose. The best 10 local regions selected from all the diagonals put together are then saved.

- Rescan the regions taken using the scoring matrices. Trimming the ends of the region to include only those contributing to the highest score.

Rescan the 10 regions taken. This time use the relevant scoring matrix while rescoring to allow runs of identities shorter than the kmer value. Also while rescoring conservative replacements that contribute to the similarity score are taken. Though protein sequences use the BLOSUM50 matrix, scoring matrices based on the minimum number of base changes required for a specific replacement, on identities alone, or on an alternative measure of similarity such as PAM, can also be used with the program. For each of the diagonal regions rescanned this way, a subregion with the maximum score is identified. The initial scores found in step1 are used to rank the library sequences. The highest score is referred to as init1 score.

- In an alignment if several initial regions with scores greater than a CUTOFF value are found, check whether the trimmed initial regions can be joined to form an approximate alignment with gaps. Calculate a similarity score that is the sum of the joined regions penalising for each gap 20 points. This initial similarity score (initn) is used to rank the library sequences. The score of the single best initial region found in step 2 is reported (init1).

Here the program calculates an optimal alignment of initial regions as a combination of compatible regions with maximal score. This optimal alignment of initial regions can be rapidly calculated using a dynamic programming algorithm. The resulting score initn is used to rank the library sequences. This joining process increases sensitivity but decreases selectivity. A carefully calculated cut-off value is thus used to control where this step is implemented, a value that is approximately one standard deviation above the average score expected from unrelated sequences in the library. A 200-residue query sequence with kmer 2 uses a value 28.

- Use a banded Smith–Waterman algorithm to calculate an optimal score for alignment.

This step uses a banded Smith–Waterman algorithm to create an optimised score (opt) for each alignment of query sequence to a database(library) sequence. It takes a band of 32 residues centered on the init1 region of step2 for calculating the optimal alignment. After all sequences are searched the program plots the initial scores of each database sequence in a histogram, and calculates the statistical significance of the "opt" score. For protein sequences, the final alignment is produced using a full Smith–Waterman alignment. For DNA sequences, a banded alignment is provided.

FASTA cannot remove low complexity regions before aligning the sequences as it is possible with BLAST. This might be problematic as when the query sequence contains such regions, e.g. mini- or microsatellites repeating the same short sequence frequent times, this increases the score of not familiar sequences in the database which only match in this repeats, which occur quite frequently. Therefore, the program PRSS is added in the FASTA distribution package. PRSS shuffles the matching sequences in the database either on the one-letter level or it shuffles short segments which length the user can determine. The shuffled sequences are now aligned again and if the score is still higher than expected this is caused by the low complexity regions being mixed up still mapping to the query. By the amount of the score the shuffled sequences still attain PRSS now can predict the significance of the score of the original sequences. The higher the score of the shuffled sequences the less significant the matches found between original database and query sequence.

The FASTA programs find regions of local or global similarity between Protein or DNA sequences, either by searching Protein or DNA databases, or by identifying local duplications within a sequence. Other programs provide information on the statistical significance of an alignment. Like BLAST, FASTA can be used to infer functional and evolutionary relationships between sequences as well as help identify members of gene families.

BLAST

In bioinformatics, BLAST (basic local alignment search tool) is an algorithm and program for comparing primary biological sequence information, such as the amino-acid sequences of proteins or the nucleotides of DNA and RNA sequences. A BLAST search enables a researcher to compare a subject protein or nucleotide sequence (called a query) with a library or database of sequences, and identify library sequences that resemble the query sequence above a certain threshold.

Different types of BLASTs are available according to the query sequences and the target databases. For example, following the discovery of a previously unknown gene in the mouse, a scientist will typically perform a BLAST search of the human genome to see if humans carry a similar gene; BLAST will identify sequences in the human genome that resemble the mouse gene based on similarity of sequence.

BLAST, which The New York Times called the Google of biological research, is one of the most widely used bioinformatics programs for sequence searching. It addresses a fundamental problem in bioinformatics research. The heuristic algorithm it uses is much faster than other approaches, such as calculating an optimal alignment. This emphasis on speed is vital to making the algorithm practical on the huge genome databases currently available, although subsequent algorithms can be even faster.

Before BLAST, FASTA was developed by David J. Lipman and William R. Pearson in 1985. Before fast algorithms such as BLAST and FASTA were developed, searching databases for protein or nucleic sequences was very time consuming because a full alignment procedure (e.g., the Smith–Waterman algorithm) was used.

BLAST came from the 1990 stochastic model of Samuel Karlin and Stephen Altschul They "proposed a method for estimating similarities between the known DNA sequence of one organism with that of another," and their work has been described as "the statistical foundation for BLAST." Subsequently, Altschul, along with Warren Gish, Webb Miller, Eugene Myers, and David J. Lipman at the National Institutes of Health designed the BLAST algorithm.

While BLAST is faster than any Smith-Waterman implementation for most cases, it cannot "guarantee the optimal alignments of the query and database sequences" as Smith-Waterman algorithm does. The optimality of Smith-Waterman "ensured the best performance on accuracy and the most precise results" at the expense of time and computer power.

BLAST is more time-efficient than FASTA by searching only for the more significant patterns in the sequences, yet with comparative sensitivity. This could be further realized by understanding the algorithm of BLAST.

Examples of other questions that researchers use BLAST to answer are:

- Which bacterial species have a protein that is related in lineage to a certain protein with known amino-acid sequence.
- What other genes encode proteins that exhibit structures or motifs such as ones that have just been determined.

BLAST is also often used as part of other algorithms that require approximate sequence matching. BLAST is available on the web on the NCBI website. Alternative implementations include AB-BLAST (formerly known as WU-BLAST), FSA-BLAST, and ScalaBLAST.

Input

Input sequences (in FASTA or Genbank format) and weight matrix.

Output

BLAST output can be delivered in a variety of formats. These formats include HTML, plain text, and XML formatting. For NCBI's web-page, the default format for output is HTML. When performing a BLAST on NCBI, the results are given in a graphical format showing the hits found, a table showing sequence identifiers for the hits with scoring related data, as well as alignments for the sequence of interest and the hits received with corresponding BLAST scores for these. The easiest to read and most informative of these is probably the table.

If one is attempting to search for a proprietary sequence or simply one that is unavailable in databases available to the general public through sources such as NCBI, there is a BLAST program available for download to any computer, at no cost. This can be found at BLAST+ executables.

Process

Using a heuristic method, BLAST finds similar sequences, by locating short matches between the two sequences. This process of finding similar sequences is called seeding. It is after this first match that BLAST begins to make local alignments. While attempting to find similarity in sequences, sets of common letters, known as words, are very important. For example, suppose that the sequence contains the following stretch of letters, GLKFA. If a BLAST was being conducted under normal conditions, the word size would be 3 letters. In this case, using the given stretch of letters, the searched words would be GLK, LKF, KFA. The heuristic algorithm of BLAST locates all common three-letter words between the sequence of interest and the hit sequence or sequences from the database. This result will then be used to build an alignment. After making words for the sequence of interest, the rest of the words are also assembled. These words must satisfy a requirement of having a score of at least the threshold T, when compared by using a scoring matrix.

One commonly used scoring matrix for BLAST searches is BLOSUM62, although the optimal scoring matrix depends on sequence similarity. Once both words and neighborhood words are assembled and compiled, they are compared to the sequences in the database in order to find matches. The threshold score T determines whether or not a particular word will be included in the alignment. Once seeding has been conducted, the alignment which is only 3 residues long, is extended in both directions by the algorithm used by BLAST. Each extension impacts the score of the alignment by either increasing or decreasing it. If this score is higher than a pre-determined T, the alignment will be included in the results given by BLAST. However, if this score is lower than this pre-determined T, the alignment will cease to extend, preventing the areas of poor alignment from being included in the BLAST results. Note that increasing the T score limits the amount of space available to search, decreasing the number of neighborhood words, while at the same time speeding up the process of BLAST.

Algorithm

To run the software, BLAST requires a query sequence to search for, and a sequence to search against (also called the target sequence) or a sequence database containing multiple such sequences. BLAST will find sub-sequences in the database which are similar to sub sequences in the query. In typical usage, the query sequence is much smaller than the database, e.g., the query may be one thousand nucleotides while the database is several billion nucleotides.

The main idea of BLAST is that there are often High-scoring Segment Pairs (HSP) contained in a statistically significant alignment. BLAST searches for high scoring sequence alignments between the query sequence and the existing sequences in the database using a heuristic approach that approximates the Smith-Waterman algorithm. However, the exhaustive Smith-Waterman approach is too slow for searching large genomic databases such as GenBank. Therefore, the BLAST algorithm uses a heuristic approach that is less accurate than the Smith-Waterman algorithm but over 50 times faster. The speed and relatively good accuracy of BLAST are among the key technical innovations of the BLAST programs.

An overview of the BLAST algorithm (a protein to protein search) is as follows:

- Remove low-complexity region or sequence repeats in the query sequence: "Low-complexity region" means a region of a sequence composed of few kinds of elements. These regions might give high scores that confuse the program to find the actual significant sequences in the database, so they should be filtered out. The regions will be marked with an X (protein sequences) or N (nucleic acid sequences) and then be ignored by the BLAST program. To filter out the low-complexity regions, the SEG program is used for protein sequences and the program DUST is used for DNA sequences. On the other hand, the program XNU is used to mask off the tandem repeats in protein sequences.

- Make a k-letter word list of the query sequence: Take k=3 for example, we list the words of length 3 in the query protein sequence (k is usually 11 for a DNA sequence) "sequentially", until the last letter of the query sequence is included. The method is illustrated in figure.

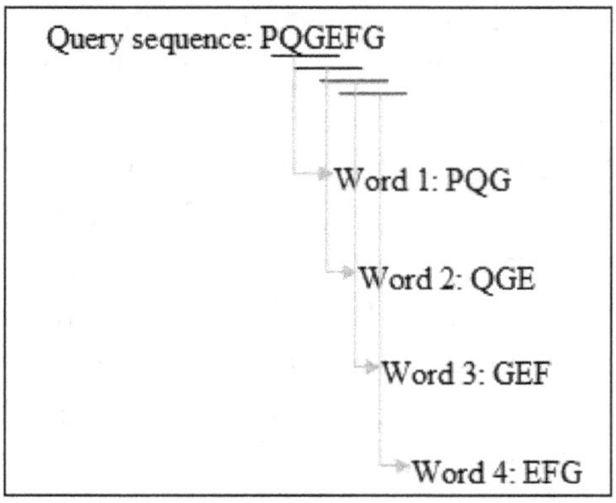

The method to establish the k-letter query word list.

- List the possible matching words: This step is one of the main differences between BLAST and FASTA. FASTA cares about all of the common words in the database and query sequences that are listed in the above step; however, BLAST only cares about the high-scoring words. The scores are created by comparing the word in the list in the above step with all the 3-letter words. By using the scoring matrix (substitution matrix) to score the comparison of each residue pair, there are 20^3 possible match scores for a 3-letter word. For example, the score obtained by comparing PQG with PEG and PQA is respectively 15 and 12 with the BLOSUM62 weighting scheme. For DNA words, a match is scored as +5 and a mismatch as -4, or as +2 and -3. After that, a neighborhood word score threshold T is used to reduce the number of possible matching words. The words whose scores are greater than the threshold T will remain in the possible matching words list, while those with lower scores will be discarded. For example, PEG is kept, but PQA is abandoned when T is 13.

- Organize the remaining high-scoring words into an efficient search tree: This allows the program to rapidly compare the high-scoring words to the database sequences.

- Repeat step 3 to 4 for each k-letter word in the query sequence.

- Scan the database sequences for exact matches with the remaining high-scoring words: The BLAST program scans the database sequences for the remaining high-scoring word, such as PEG, of each position. If an exact match is found, this match is used to seed a possible un-gapped alignment between the query and database sequences.

- Extend the exact matches to high-scoring segment pair (HSP):

    - The original version of BLAST stretches a longer alignment between the query and the database sequence in the left and right directions, from the position where the exact match occurred. The extension does not stop until the accumulated total score of the HSP begins to decrease. A simplified example is presented in figure.

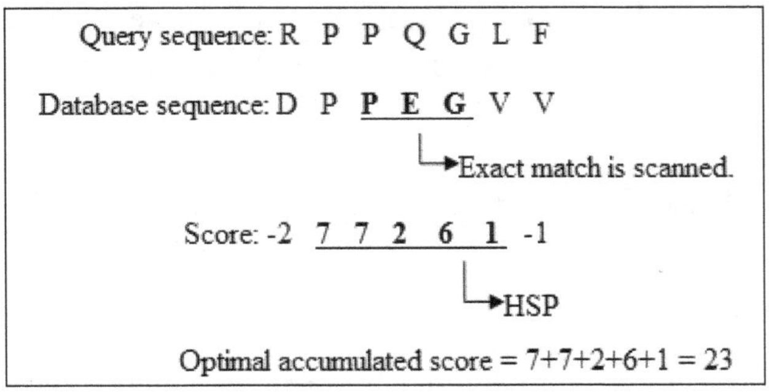

The process to extend the exact match. Adapted from Biological Sequence Analysis I, Current Topics in Genome Analysis.

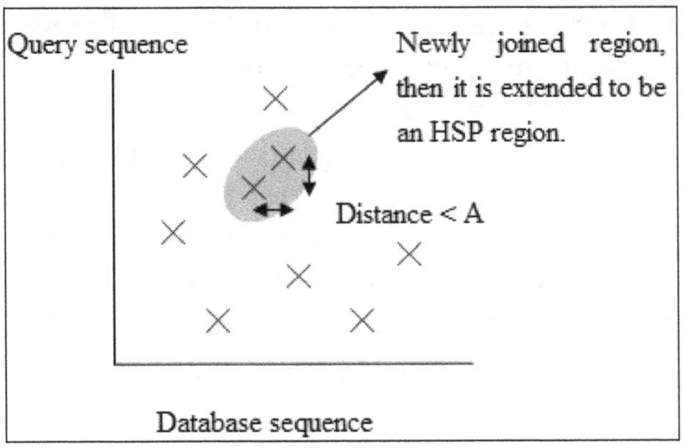

The positions of the exact matches.

- To save more time, a newer version of BLAST, called BLAST2 or gapped BLAST, has been developed. BLAST2 adopts a lower neighborhood word score threshold to maintain the same level of sensitivity for detecting sequence similarity. Therefore, the possible matching words list in step 3 becomes longer. Next, the exact matched regions, within distance A from each other on the same diagonal in figure will be joined as a longer new region. Finally, the new regions are then extended by the same method as in the original version of BLAST, and the HSPs' (High-scoring segment pair) scores of the extended regions are then created by using a substitution matrix as before.

- List all of the HSPs in the database whose score is high enough to be considered: We list the HSPs whose scores are greater than the empirically determined cutoff score S. By examining the distribution of the alignment scores modeled by comparing random sequences, a cutoff score S can be determined such that its value is large enough to guarantee the significance of the remaining HSPs.
- Evaluate the significance of the HSP score: BLAST next assesses the statistical significance of each HSP score by exploiting the Gumbel extreme value distribution (EVD). (It is proved that the distribution of Smith-Waterman local alignment scores between two random sequences follows the Gumbel EVD. For local

alignments containing gaps it is not proved.). In accordance with the Gumbel EVD, the probability p of observing a score S equal to or greater than x is given by the equation:

$$p(S \geq x) = 1 - \exp\left(-e^{-\lambda(x-\mu)}\right),$$

where,

$$\mu = \frac{\log(Km'n')}{\lambda} .$$

The statistical parameters λ and K are estimated by fitting the distribution of the un-gapped local alignment scores, of the query sequence and a lot of shuffled versions (Global or local shuffling) of a database sequence, to the Gumbel extreme value distribution. Note that λ and K depend upon the substitution matrix, gap penalties, and sequence composition (the letter frequencies). m' and n' are the effective lengths of the query and database sequences, respectively. The original sequence length is shortened to the effective length to compensate for the edge effect (an alignment start near the end of one of the query or database sequence is likely not to have enough sequence to build an optimal alignment). They can be calculated as:

$$m' \approx m - \frac{\ln Kmn}{H} ,$$

$$n' \approx n - \frac{\ln Kmn}{H} .$$

where H is the average expected score per aligned pair of residues in an alignment of two random sequences. Altschul and Gish gave the typical values, $\lambda = 0.318$, $K = 0.13$, and $H = 0.40$, for un-gapped local alignment using BLOSUM62 as the substitution matrix. Using the typical values for assessing the significance is called the lookup table method; it is not accurate. The expect score E of a database match is the number of times that an unrelated database sequence would obtain a score S higher than x by chance. The expectation E obtained in a search for a database of D sequences is given by:

$$E \approx 1 - e^{-p(s>x)D} .$$

Furthermore, when $p < 0.1$, E could be approximated by the Poisson distribution as:

$$E \approx pD .$$

This expectation or expect value "E" (often called an E score or E-value or e-value) assessing the significance of the HSP score for un-gapped local alignment is reported in the BLAST results. The calculation shown here is modified if individual HSPs are combined, such as when producing gapped alignments (described below), due to the variation of the statistical parameters.

- Make two or more HSP regions into a longer alignment: Sometimes, we find two or more HSP regions in one database sequence that can be made into a longer alignment. This provides additional evidence of the relation between the query and database sequence. There are two methods, the Poisson method and the sum-of-scores method, to compare the significance of the newly combined HSP regions. Suppose that there are two combined HSP regions with the pairs of scores (65, 40) and (52, 45), respectively. The Poisson method gives more significance to the set with the maximal lower score (45>40). However, the sum-of-scores method prefers the first set, because 65+40 (105) is greater than 52+45(97). The original BLAST uses the Poisson method; gapped BLAST and the WU-BLAST uses the sum-of scores method.
- Show the gapped Smith-Waterman local alignments of the query and each of the matched database sequences.
    - The original BLAST only generates un-gapped alignments including the initially found HSPs individually, even when there is more than one HSP found in one database sequence.
    - BLAST2 produces a single alignment with gaps that can include all of the initially found HSP regions. Note that the computation of the score and its corresponding E-value involves use of adequate gap penalties.
- Report every match whose expect score is lower than a threshold parameter E.

Parallel BLAST

Parallel BLAST versions of split databases are implemented using MPI and Pthreads, and have been ported to various platforms including Windows, Linux, Solaris, Mac OS X, and AIX. Popular approaches to parallelize BLAST include query distribution, hash table segmentation, computation parallelization, and database segmentation (partition). Databases are split into equal sized pieces and stored locally on each node. Each query is run on all nodes in parallel and the resultant BLAST output files from all nodes merged to yield the final output. Specific implementations include MPIblast, ScalaBLAST, DCBLAST and so on.

Program

The BLAST program can either be downloaded and run as a command-line utility "blastall" or accessed for free over the web. The BLAST web server, hosted by the NCBI, allows anyone with a web browser to perform similarity searches against constantly updated databases of proteins and DNA that include most of the newly sequenced organisms.

The BLAST program is based on an open-source format, giving everyone access to it and enabling them to have the ability to change the program code. This has led to the creation of several BLAST "spin-offs".

There are now a handful of different BLAST programs available, which can be used depending on what one is attempting to do and what they are working with. These different programs vary in query sequence input, the database being searched, and what is being compared. These programs and their details are listed below:

BLAST is actually a family of programs (all included in the blastall executable). These include:

- Nucleotide-nucleotide BLAST (blastn): This program, given a DNA query, returns the most similar DNA sequences from the DNA database that the user specifies.
- Protein-protein BLAST (blastp): This program, given a protein query, returns the most similar protein sequences from the protein database that the user specifies.
- Position-Specific Iterative BLAST (PSI-BLAST) (blastpgp): This program is used to find distant relatives of a protein. First, a list of all closely related proteins is created. These proteins are combined into a general "profile" sequence, which summarises significant features present in these sequences. A query against the protein database is then run using this profile, and a larger group of proteins is found. This larger group is used to construct another profile, and the process is repeated.By including related proteins in the search, PSI-BLAST is much more sensitive in picking up distant evolutionary relationships than a standard protein-protein BLAST.
- Nucleotide 6-frame translation-protein (blastx): This program compares the six-frame conceptual translation products of a nucleotide query sequence (both strands) against a protein sequence database.
- Nucleotide 6-frame translation-nucleotide 6-frame translation (tblastx): This program is the slowest of the BLAST family. It translates the query nucleotide sequence in all six possible frames and compares it against the six-frame translations of a nucleotide sequence database. The purpose of tblastx is to find very distant relationships between nucleotide sequences.
- Protein-nucleotide 6-frame translation (tblastn): This program compares a protein query against the all six reading frames of a nucleotide sequence database.
- Large numbers of query sequences (megablast): When comparing large numbers of input sequences via the command-line BLAST, "megablast" is much faster than running BLAST multiple times. It concatenates many input sequences together to

form a large sequence before searching the BLAST database, then post-analyzes the search results to glean individual alignments and statistical values.

Of these programs, BLASTn and BLASTp are the most commonly used because they use direct comparisons, and do not require translations. However, since protein sequences are better conserved evolutionarily than nucleotide sequences, tBLASTn, tBLASTx, and BLASTx, produce more reliable and accurate results when dealing with coding DNA. They also enable one to be able to directly see the function of the protein sequence, since by translating the sequence of interest before searching often gives you annotated protein hits.

Alternative Versions

A version designed for comparing large genomes or DNA is BLASTZ. CS-BLAST (Context-Specific BLAST) is an extended version of BLAST for searching protein sequences that finds twice as many remotely related sequences as BLAST at the same speed and error rate. In CS-BLAST, the mutation probabilities between amino acids depend not only on the single amino acid, as in BLAST, but also on its local sequence context. Washington University produced an alternative version of NCBI BLAST, called WU-BLAST. The rights have since been acquired to Advanced Biocomputing, LLC.

In 2009, NCBI has released a new set of BLAST executables, the C++ based BLAST+, and has released C versions until 2.2.26. Starting with version 2.2.27, only BLAST+ executables are available. Among the changes is the replacement of the `blastall` executable with separate executables for the different BLAST programs, and changes in option handling. The formatdb utility (C based) has been replaced by makeblastdb (C++ based) and databases formatted by either one should be compatible for identical blast releases. The algorithms remain similar, however, the number of hits found and their order can vary significantly between the older and the newer version.

Accelerated Versions

TimeLogic offers an FPGA-accelerated implementation of the BLAST algorithm called Tera-BLAST that is hundreds of times faster.

Other formerly supported versions include:

- FPGA-accelerated:
    - Prior to their acquisition by Qiagen, CLC bio collaborated with SciEngines GmbH on an FPGA accelerator they claimed will give 188x acceleration of BLAST.
    - The Mitrion-C Open Bio Project was an effort to port BLAST to run on Mitrion FPGAs.

- GPU-accelerated:
  - GPU-Blast is an accelerated version of NCBI BLASTP for CUDA which is 3x~4x faster than NCBI Blast.
  - CUDA-BLASTP is a version of BLASTP that is GPU-accelerated and is claimed to run up to 10x faster than NCBI BLAST.
  - G-BLASTN is an accelerated version of NCBI blastn and megablast, whose speedup varies from 4x to 14x (compared to the same runs with 4 CPU threads). Its current limitation is that the database must fit into the GPU memory.
- CPU-accelerated:
  - MPIBlast is a parallel implementation of NCBI BLAST using Message Passing Interface. By efficiently utilizing distributed computational resources through database fragmentation, query segmentation, intelligent scheduling, and parallel I/O, mpiBLAST improves NCBI BLAST performance by several orders of magnitude while scaling to hundreds of processors.
  - CaBLAST makes search on large databases orders of magnitude faster by exploiting redundancy in data.
  - Paracel BLAST was a commercial parallel implementation of NCBI BLAST, supporting hundreds of processors.
  - QuickBLAST (kblastp) from NCBI is an implementation accelerated by prefiltering based on Jaccard index estimates with hashed pentameric fragments. The filtering slightly reduces sensitivity, but increases performance by an order of magnitude. NCBI only makes the search available on their non-redundant (nr) protein collection, and does not offer downloads.

Alternatives to BLAST

The predecessor to BLAST, FASTA, can also be used for protein and DNA similarity searching. FASTA provides a similar set of programs for comparing proteins to protein and DNA databases, DNA to DNA and protein databases, and includes additional programs for working with unordered short peptides and DNA sequences. In addition, the FASTA package provides SSEARCH, a vectorized implementation of the rigorous Smith-Waterman algorithm. FASTA is slower than BLAST, but provides a much wider range of scoring matrices, making it easier to tailor a search to a specific evolutionary distance.

An extremely fast but considerably less sensitive alternative to BLAST is BLAT (Blast Like Alignment Tool). While BLAST does a linear search, BLAT relies on k-mer indexing the database, and can thus often find seeds faster. Another software alternative similar to BLAT is PatternHunter.

Advances in sequencing technology in the late 2000s has made searching for very similar nucleotide matches an important problem. New alignment programs tailored for this use typically use BWT-indexing of the target database (typically a genome). Input sequences can then be mapped very quickly, and output is typically in the form of a BAM file. Example alignment programs are BWA, SOAP, and Bowtie.

For protein identification, searching for known domains (for instance from Pfam) by matching with Hidden Markov Models is a popular alternative, such as HMMER.

An alternative to BLAST for comparing two banks of sequences is PLAST. PLAST provides a high-performance general purpose bank to bank sequence similarity search tool relying on the PLAST and ORIS algorithms. Results of PLAST are very similar to BLAST, but PLAST is significantly faster and capable of comparing large sets of sequences with a small memory (i.e. RAM) footprint.

For applications in metagenomics, where the task is to compare billions of short DNA reads against tens of millions of protein references, DIAMOND runs at up to 20,000 times as fast as BLASTX, while maintaining a high level of sensitivity.

The open-source software MMseqs is an alternative to BLAST/PSI-BLAST, which improves on current search tools over the full range of speed-sensitivity trade-off, achieving sensitivities better than PSI-BLAST at more than 400 times its speed.

Optical computing approaches have been suggested as promising alternatives to the current electrical implementations. OptCAM is an example of such approaches and is shown to be faster than BLAST.

Comparing BLAST and the Smith-Waterman Process

While both Smith-Waterman and BLAST are used to find homologous sequences by searching and comparing a query sequence with those in the databases, they do have their differences.

Due to the fact that BLAST is based on a heuristic algorithm, the results received through BLAST, in terms of the hits found, may not be the best possible results, as it will not provide you with all the hits within the database. BLAST misses hard to find matches.

A better alternative in order to find the best possible results would be to use the Smith-Waterman algorithm. This method varies from the BLAST method in two areas, accuracy and speed. The Smith-Waterman option provides better accuracy, in that it finds matches that BLAST cannot, because it does not miss any information. Therefore, it is necessary for remote homology. However, when compared to BLAST, it is more time consuming, not to mention that it requires large amounts of computer usage and space. However, technologies to speed up the Smith-Waterman process have been found to improve the time necessary to perform a search dramatically. These technologies include FPGA chips and SIMD technology.

In order to receive better results from BLAST, the settings can be changed from their default settings. However, there is no given or set way of changing these settings in order to receive the best results for a given sequence. The settings available for change are E-Value, gap costs, filters, word size, and substitution matrix. Note, that the algorithm used for BLAST was developed from the algorithm used for Smith-Waterman. BLAST employs an alignment which finds "local alignments between sequences by finding short matches and from these initial matches (local) alignments are created".

BLAST Output Visualization

To help users interpreting BLAST results, different software is available. According to installation and use, analysis features and technology, here are some available tools:

- NCBI BLAST service.
- General BLAST output interpreters, GUI-based: JAMBLAST, Blast Viewer, BLASTGrabber.
- Integrated BLAST environments: PLAN, BlastStation-Free.
- BLAST output parsers: MuSeqBox, Zerg, BioParser, BLAST-Explorer.
- Specialized BLAST-related tools: MEGAN, BLAST2GENE, BOV, Circoletto.

Uses of BLAST

BLAST can be used for several purposes. These include identifying species, locating domains, establishing phylogeny, DNA mapping, and comparison.

- Identifying species: With the use of BLAST, you can possibly correctly identify a species or find homologous species. This can be useful, for example, when you are working with a DNA sequence from an unknown species.
- Locating domains: When working with a protein sequence you can input it into BLAST, to locate known domains within the sequence of interest.
- Establishing phylogeny: Using the results received through BLAST you can create a phylogenetic tree using the BLAST web-page. Phylogenies based

on BLAST alone are less reliable than other purpose-built computational phylogenetic methods, so should only be relied upon for "first pass" phylogenetic analyses.

- DNA mapping: When working with a known species, and looking to sequence a gene at an unknown location, BLAST can compare the chromosomal position of the sequence of interest, to relevant sequences in the database(s). NCBI has a "Magic-BLAST" tool built around BLAST for this purpose.
- Comparison: When working with genes, BLAST can locate common genes in two related species, and can be used to map annotations from one organism to another.

BLAT

BLAT (BLAST-like alignment tool) is a pairwise sequence alignment algorithm that was developed by Jim Kent at the University of California Santa Cruz (UCSC) in the early 2000s to assist in the assembly and annotation of the human genome. It was designed primarily to decrease the time needed to align millions of mouse genomic reads and expressed sequence tags against the human genome sequence. The alignment tools of the time were not capable of performing these operations in a manner that would allow a regular update of the human genome assembly. Compared to pre-existing tools, BLAT was ~500 times faster with performing mRNA/DNA alignments and ~50 times faster with protein/protein alignments.

BLAT is one of multiple algorithms developed for the analysis and comparison of biological sequences such as DNA, RNA and proteins, with a primary goal of inferring homology in order to discover biological function of genomic sequences. It is not guaranteed to find the mathematically optimal alignment between two sequences like the classic Needleman-Wunsch and Smith-Waterman dynamic programming algorithms do; rather, it first attempts to rapidly detect short sequences which are more likely to be homologous, and then it aligns and further extends the homologous regions. It is similar to the heuristic BLAST family of algorithms, but each tool has tried to deal with the problem of aligning biological sequences in a timely and efficient manner by attempting different algorithmic techniques.

Uses of BLAT

BLAT can be used to align DNA sequences as well as protein and translated nucleotide (mRNA or DNA) sequences. It is designed to work best on sequences with great similarity. The DNA search is most effective for primates and the protein search is effective for land vertebrates. In addition, protein or translated sequence queries are more effec tive for identifying distant matches and for cross-species analysis than DNA sequence queries. Typical uses of BLAT include the following:

- Alignment of multiple mRNA sequences onto a genome assembly in order to infer their genomic coordinates.
- Alignment of a protein or mRNA sequence from one species onto a sequence database from another species to determine homology. Provided the two species are not too divergent, cross-species alignment is generally effective with BLAT. This is possible because BLAT does not require perfect matches, but rather accepts mismatches in alignments.
- BLAT can be used for alignments of two protein sequences. However, it is not the tool of choice for these types of alignments. BLASTP, the Standard Protein BLAST tool, is more efficient at protein-protein alignments.
- Determination of the distribution of exonic and intronic regions of a gene.
- Detection of gene family members of a specific gene query.
- Display of the protein-coding sequence of a specific gene.

BLAT is designed to find matches between sequences of length at least 40 bases that share ≥95% nucleotide identity or ≥80% translated protein identity.

Process

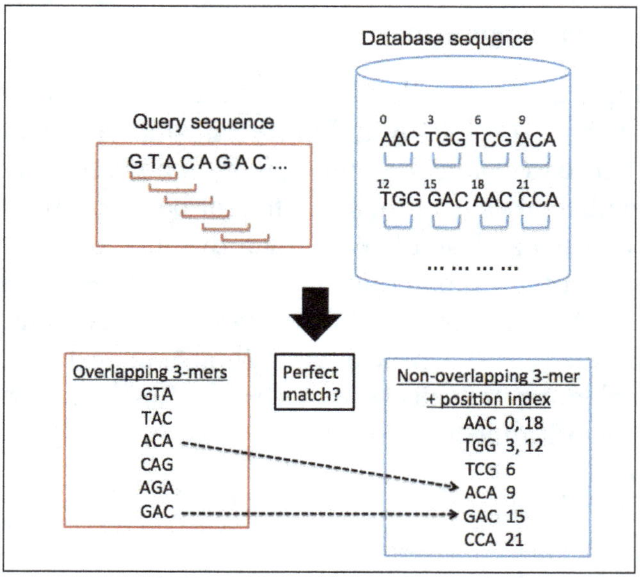

Example showing the creation of non-overlapping k-mers from the target database and overlapping k-mers from the query sequence, for k=3. Coordinates of the database sequences are used to clump the matches into larger alignments.

BLAT is used to find regions in a target genomic database which are similar to a query sequence under examination. The general algorithmic process followed by BLAT is similar to BLAST's in that it first searches for short segments in the database and query sequences which have a certain number of matching elements. These alignment seeds are then extended in both directions of the sequences in order to form high-scoring pairs. However, BLAT uses a different indexing approach from BLAST, which allows it to rapidly scan very large genomic and protein databases for similarities to a query sequence. It does this by keeping an indexed list (hash table) of the target database in memory, which significantly reduces the time required for the comparison of the query sequences with the target database. This index is built by taking the coordinates of all the non-overlapping k-mers (words with k letters) in the target database, except for highly repeated k-mers. BLAT then builds a list of all overlapping k-mers from the query sequence and searches for these in the target database, building up a list of hits where there are matches between the sequences.

Search Stage

There are three different strategies used in order to search for candidate homologous regions:

- The first method requires single perfect matches between the query and database sequences i.e. the two k-mer words are exactly the same. This approach is not considered the most practical. This is because a small k-mer size is necessary in order to achieve high levels of sensitivity, but this increases the number of false positive hits, thus increasing the amount of time spent in the alignment stage of the algorithm.
- The second method allows at least one mismatch between the two k-mer words. This decreases the amount of false positives, allowing larger k-mer sizes which are less computationally expensive to handle than those produced from the previous method. This method is very effective in identifying small homologous regions.
- The third method requires multiple perfect matches which are in close proximity to each other. As Kent shows, this is a very effective technique capable of taking into consideration small insertions and deletions within the homologous regions.

When aligning nucleotides, BLAT uses the third method requiring two perfect word matches of size 11 (11-mers). When aligning proteins, the BLAT version determines the search methodology used: when the client/server version is used, BLAT searches for three perfect 4-mer matches; when the stand-alone version is used, BLAT searches for a single perfect 5-mer between the query and database sequences.

BLAT vs. BLAST

Some of the differences between BLAT and BLAST are outlined below:

- BLAT indexes the genome/protein database, retains the index in memory, and then scans the query sequence for matches. BLAST, on the other hand, builds an index of the query sequences and searches through the database for matches. A BLAST variant called MegaBLAST indexes 4 databases to speed up alignments.
- BLAT can extend on multiple perfect and near-perfect matches (default is 2 perfect matches of length 11 for nucleotide searches and 3 perfect matches of length 4 for protein searches), while BLAST extends only when one or two matches occur close together.
- BLAT connects each homologous area between two sequences into a single larger alignment, in contrast to BLAST which returns each homologous area as a separate local alignment. The result of BLAST is a list of exons with each alignment extending just past the end of the exon. BLAT, however, correctly places each base of the mRNA onto the genome, using each base only once and can be used to identify intron-exon boundaries (i.e. splice sites).
- BLAT is less sensitive than BLAST.

Program Usage

BLAT can be used either as a web-based server-client program or as a stand-alone program.

Server-client

The web-based application of BLAT can be accessed from the UCSC Genome Bioinformatics Site. Building the index is a relatively slow procedure. Therefore, each genome assembly used by the web-based BLAT is associated with a BLAT server, in order to have a pre-computed index available for alignments. These web-based BLAT servers keep the index in memory for users to input their query sequences.

Once the query sequence is uploaded/pasted into the search field, the user can select various parameters such as which species' genome to target (there are currently over 50 species available) and the assembly version of that genome (for example, the human genome has four assemblies to select from), the query type (i.e. whether the sequence relates to DNA, protein etc.) and output settings (i.e. how to sort and visualise the output). The user can then run the search by either submitting the query or using the BLAT "I'm feeling lucky" search.

Bhagwat et al. provide step by step protocols for how to use BLAT to:

- Map an mRNA/cDNA sequence to a genomic sequence.
- Map a protein sequence to the genome.
- Perform homology searches.

Input

BLAT can handle long database sequences, however, it is more effective with short query sequences than long query sequences. Kent recommends a maximum query length of 200,000 bases. The UCSC browser limits query sequences to less than 25,000 letters (i.e. nucleotides) for DNA searches and less than 10,000 letters (i.e. amino acids) for protein and translated sequence searches.

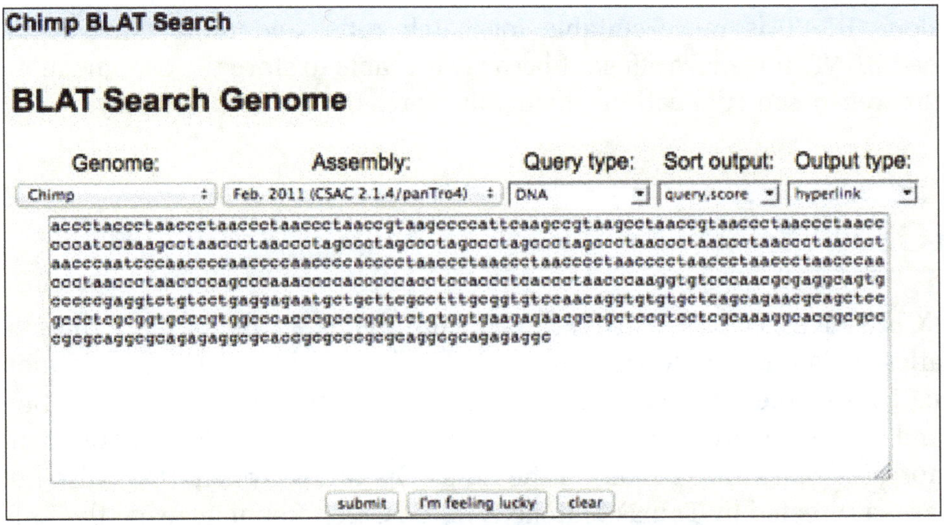

Using web-based BLAT to search a target database with a DNA query sequence. The search parameters can be seen above the query sequence.

The BLAT Search Genome available on the UCSC website accepts query sequences as text (cut and pasted into the query box) or uploaded as text files. The BLAT Search Genome can accept multiple sequences of the same type at once, up to a maximum of 25. For multiple sequences, the total number of nucleotides must not exceed 50,000 for DNA searches or 25,000 letters for protein or translated sequence searches. An example of searching a target database with a DNA query sequence is shown in figure.

Output

A BLAT search returns a list of results that are ordered in decreasing order based on the score. The following information is returned: the score of the alignment, the region of query sequence that matches to the database sequence, the size of the query sequence, the level of identity as a percentage of the alignment and the chromosome and position that the query sequence maps to. Bhagwat et al. describe how the BLAT "Score" and "Identity" measures are calculated.

For each search result, the user is provided with a link to the UCSC Genome Browser so they can visualise the alignment on the chromosome. This a major benefit of the web-based BLAT over the stand-alone BLAT. The user is able to obtain biological information associated with the alignment, such as information about the gene to which the query may match. The user is also provided with a link to view the alignment of the query sequence with the genome assembly. The matches between the query and genome assembly are blue and the boundaries of the alignments are lighter in colour. These exon boundaries indicate splice sites. The "I'm feeling lucky" search result returns the highest scoring alignment for the first query sequence based on the output sort option selected by the user.

Stand-alone

Stand-alone BLAT is more suitable for batch runs, and more efficient than the web-based BLAT. It is more efficient because it is able to store the genome in memory, unlike the web-based application which only stores the index in memory.

BIOSLAX

BioSLAX is a Live CD/Live DVD/Live USB comprising a suite of more than 300 bioinformatics tools and application suites. It has been released by the Bioinformatics Resource Unit of the Life Sciences Institute (LSI), National University of Singapore (NUS) and is bootable from any PC that allows a CD/DVD or USB boot option and runs the compressed Slackware flavour of the Linux Operating System (OS), also known as Slax. Slax was created by Tomáš Matějíček in the Czech Republic using the Linux Live Scripts which he also developed. The BioSLAX derivative was created by Mark De Silva, Lim Kuan Siong and Tan Tin Wee. BioSLAX was first released to the NUS Life Science Curriculum in April 2006.

Modularity

Because Slax worked by overlaying "application modules" on top of the base Linux OS, it made the entire distribution modular. The additional functionality of deploying these modules even while the system was already running, made using Slax even more appealing. The inclusion of the GUI based "BioSLAX Module Manager", made this process of dynamically adding and removing modules even easier.

Users were able to test updates to software or new versions and "rollback" to previous versions if they want. This was especially effective if SLAX/BioSLAX was installed to a writable medium such as a USB drive.

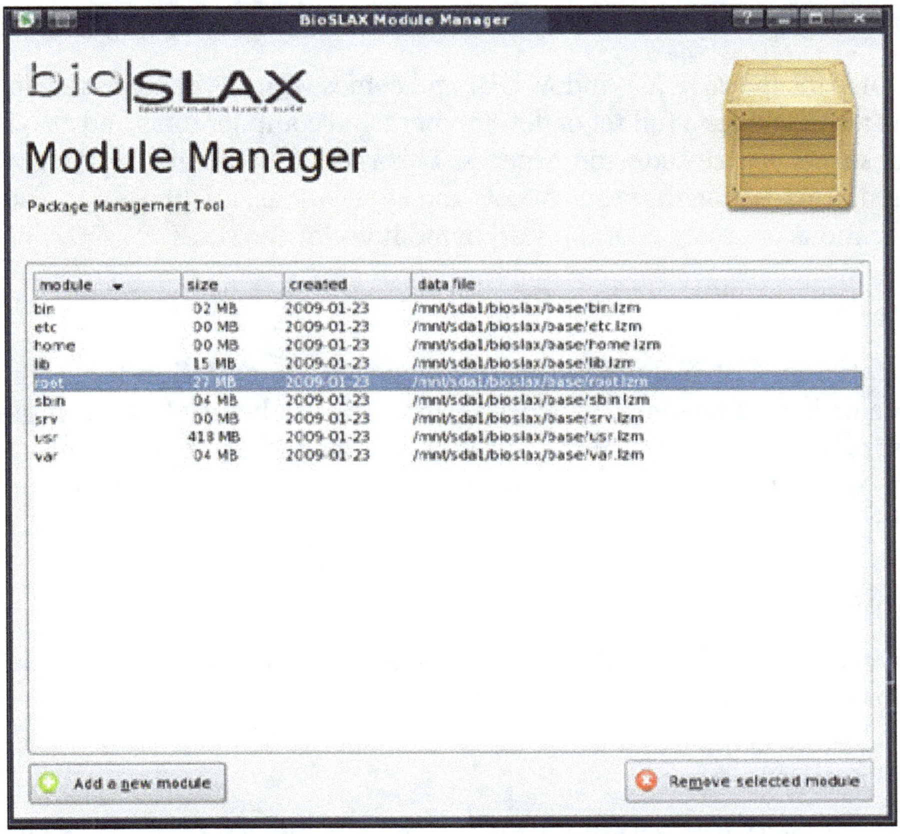

Versions

To date, there have been two versions of BioSLAX - BioSLAX 5.x based on Slax 5 and BioSLAX 7.x based on Slax 6. While BioSLAX 5.x followed the version numbers of Slax 5, BioSLAX 7 adopted a new version numbering which is one higher than the Slax version on which it is based. Latest versions can be downloaded from the BioSLAX website.

BioSLAX 5.x

BioSLAX 5.x was largely based on the 5.1.8 version of Slax, running earlier versions of the 2.6 Linux kernel and KDE 3.4, with unionfs.

BioSLAX 5.x Editions

Standard user Edition

This edition runs the KDE X Window GUI and comes with all the tools and application suites, but does not include any compiler tools nor the Linux kernel source code and headers. This is mainly suited for users who only need to use the tools and applications suites. It has a very small size, making it easy to download and particularly convenient for regions where internet bandwidth is an issue.

Developer Edition

This edition runs the KDE X Window GUI and comes with all the tools and application suites and also includes a full set of development and compiler tools and also including the Linux kernel source code and headers. This is edition is more for the power user, who, in addition to using the various tools and applications, might want to also compile new applications or create new application modules for BioSLAX.

Server Edition

This edition does not include any X Window GUI, compilation tools, Linux kernel source or kernel headers. It is primarily meant to be used as a remote server, where users have to either SSH in to use the command line applications or connect to the server via the web to access the available web-based portals to popular bio applications.

NUS LSM Edition

This edition is the Developer Edition, customized for use by the NUS Life Science Curriculum for the teaching of bioinformatics.

Taverna Edition

This edition is the Developer Edition which includes TaveRNA. The TaveRNA Project aims to provide a language and software tools to facilitate easy use of workflow and distributed compute technology.

BioSLAX 7.x

BioSLAX 7.x is based on Slax 6 and features the later releases of the 2.6 Linux kernel, KDE 3.5 and using aufs and lzma compression. The biggest change is the use of this version as either client or server. The distribution was also moved from CD to DVD, allowing for more applications to be introduced, which were previously left out of version 5.x due to space considerations. The ability to boot from a FAT or EXT formatted USB drive was also introduced in Slax 6, hence BioSLAX 7.x versions also had this feature, effectively enabling persistent file handling which are unavailable on the CD/DVD as they are not re-writable.

BioSLAX 8

Versions of BioSLAX after 7.x have been delayed due to the base distribution's (Slax) developer, Tomáš Matějíček, refusing to move forward with a new version because of family commitments. However his primary reason for not moving forward was that he was waiting for Squash FS and LZMA to be integrated into the Linux kernel by default, instead of users needing to apply separate patches. As of kernel 2.6.38, the integration was finally done and this has prompted Tomáš Matějíček to look at a new version of Slax, which will therefore result in a new version of BioSLAX in the coming months. One can follow his thoughts on the new version of Slax on his blog.

Features

Standard Tools

BioSLAX features the Linux Slackware 12.1 operating system with updated drivers for various network adapters including support for a large variety of wireless cards. It also has many useful basic tools and applications such as:

- PERL (including BioPerl modules).
- PHP.
- Apache 2.
- MySQL.
- OpenOffice.org.
- KPDF Reader.
- Mozilla Firefox.
- Mozilla Thunderbird.
- gFTP.
- ProFTPd.
- Open SSH.
- Kopete Instant Messenger.
- VNC Viewer.
- Remote Desktop Services.

BioInformatics Tools

The bioinformatics tools and applications are subdivided into three main categories.

Console Apps

- BLAST.
- BlastCL3.

- BioGrep.
- ClustalW.
- EMBOSS.
- Genesplicer.
- GlimmerHMM.
- HMMER.
- Modeller.
- PamL.
- Phylip.
- Primer3.
- R programming language and Bioconductor.
- T-Coffee.

Desktop Apps

- ACT.
- Artemis.
- ClustalX (GUI Based ClustalW).
- JAligner.
- Jalview.
- jEMBOSS (Java EMBOSS Suite).
- Jmol.
- NJPlot.
- Pymol.
- ReadSEQ.
- TreeView.
- Weka (machine learning).

Web Apps

- Web BLAST.
- Web ClustalW.
- Web Phylip.
- Web T-Coffee.
- wEMBOSS (Web based EMBOSS suite).
- Sequence Manipulation Suite (SMS).

Installing to Hard Disk

One of the more intriguing features of Slax-based distributions is how easy it is to convert the live OS into a full-fledged Linux system installed on the hard drive of any PC, which will take up roughly 3.5GBytes of space.

A tool, written with the KDE Kommander toolkit called the "BioSLAX Installer" is provided for users to easily convert their live OS to a full Linux installation. By using modules to customize the distribution and then using the installer, users can do rapid deployment of fully installed customized clients.

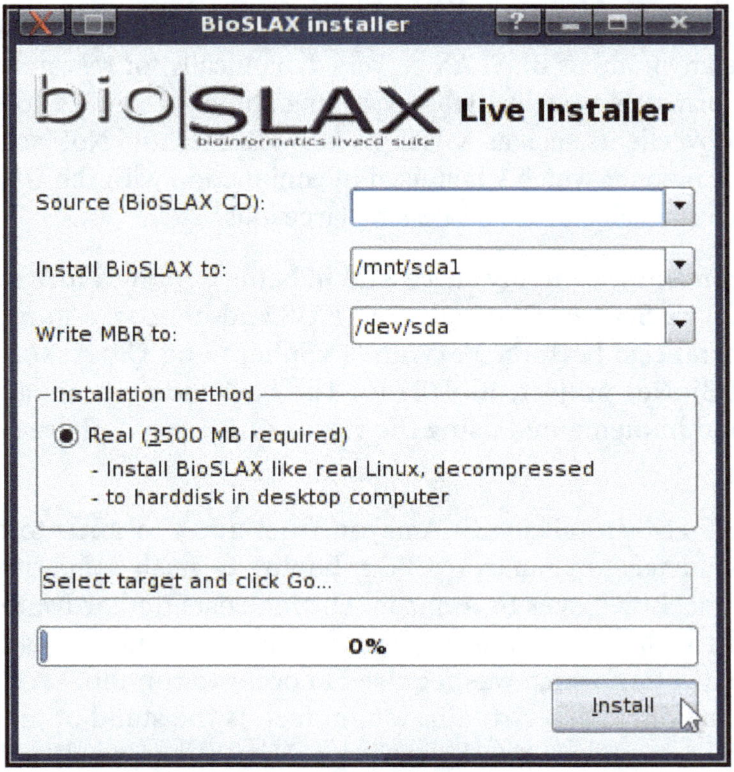

BioSLAX Updates

BioSLAX will be updated as newer Slackware (or Slax) versions are released. The tools and applications suites will also be monitored for significant changes and upgraded as necessary. Some tools may be removed to make way for other tools which can do the same thing but with added functionality and better efficiency. More web-based portals are being looked at, for example, portals to ReadSeq, Primer3 and Genesplicer are in the pipeline.

Grid Deployment

The developers were also looking at integrating various Grid computing platforms with BioSLAX. Because BioSLAX can be booted up immediately from any CD/DVD/USB, it can be used as a rapidly deployable Grid-enabled Operating System. One such Grid platform was the Univa Grid platform. Using the Univa Grid MP agent, it was shown during GridAsia 2009 in a talk given by Tan Tin Wee, that the agent, once modularized on BioSLAX, can be used to Grid enable machines from any location as slave-nodes to a master-node located elsewhere, effectively creating a "global-wide grid".

BioSLAX on the CLOUD

In a proof-of-concept endeavour, the developers successfully deployed BioSLAX as instances on a pool of resources using both VMWare's ESXi and Citrix Xen's Hypervisors. Their aim was to effectively create a "BioSLAX CLOUD" where students and staff may instantiate any number BioSLAX servers dynamically for research and education (conduct bioinformatics practical labs by having students connect to the servers via suitable X Window clients such as X-Win32, VNC, Exceed and NoMachine NX) or deployed in such a manner which when used in conjunction with the UD Grid mpagent may be used to form a cluster for processing large jobs.

The proof-of-concept was highly successful in being deployed for research and education for the Life Science Curriculum at NUS and in 2011, a number of the BioSLAX cloud instances, both on VMWare's vSphere and Citrix Xen servers, were used in the APBioNet project, BioDB100. The backend controls and automation were created and implemented using the various APIs for vSphere and Xen by Mr. Mark De Silva.

Developers were also in talks with Amazon from 2009 to 2010 to deploy similar BioSLAX cloud images on Amazon's EC2, hoping to push some of their research and education machines over to Amazon, cutting costs on hardware. Discussions, however, fell through when it was clear that Amazon was not going to support full hardware virtualization which was required in order to run BioSLAX images on the cloud. Supporting only para-virtualizaion, in fact, is the stand of most commercial cloud providers using Citrix Xen hypervisors. Until the mind-set of these entities change, only private clouds running Citrix Xen hypervisors configured for full hard ware virtualization or VMWare vSphere clouds will be the only clouds capable of running BioSLAX.

BIOPERL

BioPerl is a collection of Perl modules that facilitate the development of Perl scripts for bioinformatics applications. It has played an integral role in the Human Genome Project.

Features and Examples

BioPerl provides software modules for many of the typical tasks of bioinformatics programming. These include:

- Accessing nucleotide and peptide sequence data from local and remote databases.

Example of accessing GenBank to retrieve a sequence:

```
use Bio::DB::GenBank;


$db_obj = Bio::DB::GenBank->new;


$seq_obj = $db_obj->get_Seq_by_acc( # Insert Accession Number );
```

- Transforming formats of database/file records.

Example code for transforming formats:

```
use Bio::SeqIO;


my $usage = "all2y.pl informat outfile outfileformat";

my $informat = shift or die $usage;

my $outfile = shift or die $usage;

my $outformat = shift or die $usage;


my $seqin = Bio::SeqIO->new( -fh => *STDIN, -format => $informat, );
```

```
my $seqout = Bio::SeqIO->new( -file => ">$outfile", -format => $outfor-
mat, );


while (my $inseq = $seqin->next_seq)

{

 $seqout->write_seq($inseq);

}
```

- Manipulating individual sequences.

Example of gathering statistics for a given sequence:

```
use Bio::Tools::SeqStats;

$seq_stats = Bio::Tools::SeqStats->new($seqobj);


$weight = $seq_stats->get_mol_wt();

$monomer_ref = $seq_stats->count_monomers();


# for nucleic acid sequence

$codon_ref = $seq_stats->count_codons();
```

- Searching for similar sequences.
- Creating and manipulating sequence alignments.
- Searching for genes and other structures on genomic DNA.
- Developing machine readable sequence annotations.

Usage

In addition to being used directly by end-users, BioPerl has also provided the base for a wide variety of bioinformatic tools, including amongst others:

- SynBrowse.
- GeneComber.
- TFBS.

- MIMOX.
- BioParser.
- Degenerate primer design.
- Querying the public databases.
- Current Comparative Table.

New tools and algorithms from external developers are often integrated directly into BioPerl itself:

- Dealing with phylogenetic trees and nested taxa.
- FPC Web tools.

Advantages

BioPerl was one of the first biological module repositories that increased its usability. It has very easy to install modules, along with a flexible global repository. BioPerl uses good test modules for a large variety of processes.

Disadvantages

There are many ways to use BioPerl, from simple scripting to very complex object programming. This makes the language not clear and sometimes hard to understand. For as many modules that BioPerl has, some do not always work the way they are intended.

BIOJAVA

BioJava is an open-source software project dedicated to provide Java tools to process biological data. BioJava is a set of library functions written in the programming language Java for manipulating sequences, protein structures, file parsers, Common Object Request Broker Architecture (CORBA) interoperability, Distributed Annotation System (DAS), access to AceDB, dynamic programming, and simple statistical routines. BioJava supports a huge range of data, starting from DNA and protein sequences to the level of 3D protein structures. The BioJava libraries are useful for automating many daily and mundane bioinformatics tasks such as to parsing a Protein Data Bank (PDB) file, interacting with Jmol and many more. This application programming interface (API) provides various file parsers, data models and algorithms to facilitate working with the standard data formats and enables rapid application development and analysis.

Additional projects from BioJava include rcsb-sequenceviewer, biojava-http, biojava-spark, and rcsb-viewers.

Features

BioJava provides software modules for many of the typical tasks of bioinformatics programming. These include:

- Accessing nucleotide and peptide sequence data from local and remote databases.
- Transforming formats of database/file records.
- Protein structure parsing and manipulation.
- Manipulating individual sequences.
- Searching for similar sequences.
- Creating and manipulating sequence alignments.

Modules

BioJava 3 is a clear departure from the version 1 series. It now consists of several independent modules built using an automation tool called Apache Maven. These modules provide state-of-the-art tools for protein structure comparison, pairwise and multiple sequence alignments, working with DNA and protein sequences, analysis of amino acid properties, detecting protein modifications, predicting disordered regions in proteins, and parsers for common file formats using a biologically meaningful data model. The original code has been moved into a separate BioJava legacy project, which is still available for backward compatibility.

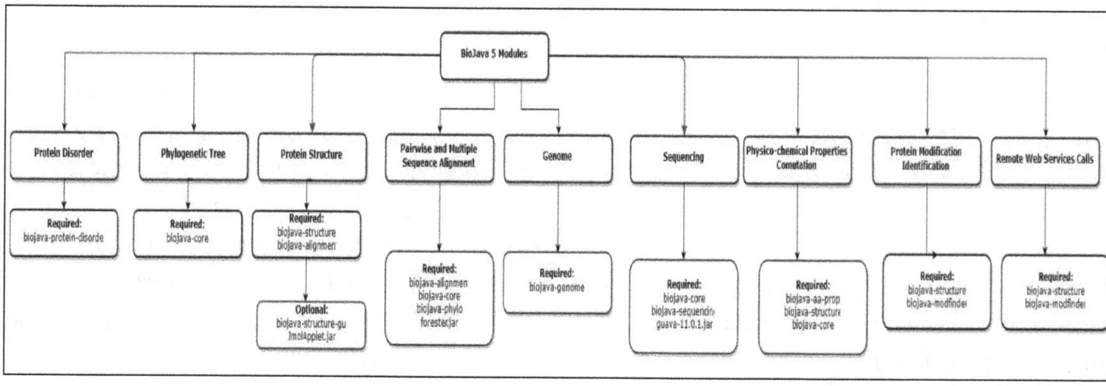

BioJava 5 introduced new features to two modules, biojava-alignment and biojava-structure. The following sections will describe several of the new modules and highlight some of the new features that are included in the latest version of BioJava.

Core Module

This module provides Java classes to model amino acid or nucleotide sequences. The classes were designed so that the names are familiar and make sense to biologists and also provide a concrete representation of the steps in going from a gene sequence to a protein sequence for computer scientists and programmers.

A major change between the legacy BioJava project and BioJava3 lies in the way framework has been designed to exploit then-new innovations in Java. A sequence is defined as a generic interface allowing the rest of the modules to create any utility that operates on all sequences. Specific classes for common sequences such as DNA and proteins have been defined in order to improve usability for biologists. The translation engine really leverages this work by allowing conversions between DNA, RNA and amino acid sequences. This engine can handle details such as choosing the codon table, converting start codons to methionine, trimming stop codons, specifying the reading frame and handing ambiguous sequences.

Special attention has been paid to designing the storage of sequences to minimize space needs. Special design patterns such as the Proxy pattern allowed the developers to create the framework such that sequences can be stored in memory, fetched on demand from a web service such as UniProt, or read from a FASTA file as needed. The latter two approaches save memory by not loading sequence data until it is referenced in the application. This concept can be extended to handle very large genomic datasets, such as NCBI GenBank or a proprietary database.

Protein Structure Modules

The protein structure modules provide tools to represent and manipulate 3D biomolecular structures. They focus on protein structure comparison.

The following algorithms have been implemented and included in BioJava:

- FATCAT algorithm for flexible and rigid body alignment.
- The standard Combinatorial Extension (CE) algorithm.
- A new version of CE that can detect circular permutations in proteins.

These algorithms are used to provide the RCSB Protein Data Bank (PDB) Protein Comparison Tool as well as systematic comparisons of all proteins in the PDB on a weekly basis.

Parsers for PDB and mmCIF file formats allow the loading of structure data into a reusable data model. This feature is used by the SIFTS project to map between UniProt sequences and PDB structures. Information from the RCSB PDB can be dynamically fetched without the need to manually download data. For visualization, an interface to the 3D viewer Jmol is provided.

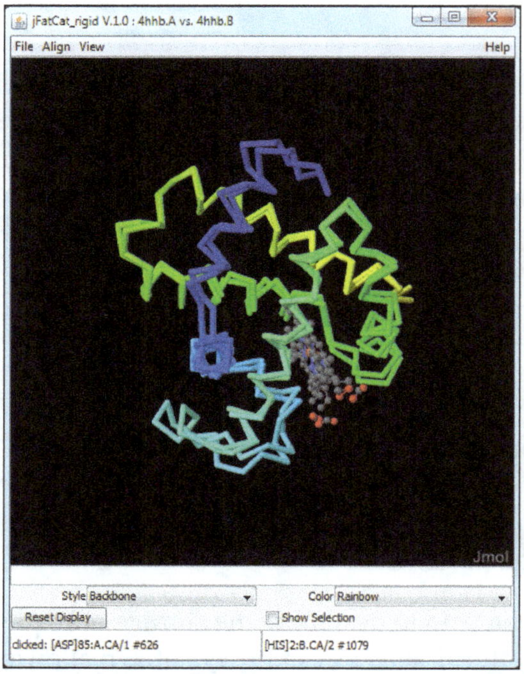

This window shows two proteins with IDs "4hhb.A" and "4hhb.B" aligned against each other. The code is given on the left side. This is produced using BioJava libraries which in turn uses Jmol viewer. The FATCAT rigid algorithm is used here to do the alignment.

Genome and Sequencing Modules

This module is focused on the creation of gene sequence objects from the core module. This is realized by supporting the parsing of the following popular standard file formats generated by open source gene prediction applications:

- GTF files generated by GeneMark.
- GFF2 files generated by GeneID.
- GFF3 files generated by Glimmer.

Then the gene sequence objects are written out as a GFF3 format and is imported into GMOD. These file formats are well defined but what gets written in the file is very flexible.

For providing input-output support for several common variants of the FASTQ file format from the next generation sequencers, a separate sequencing module is provided. For samples on how to use this module please go to this link.

Alignment Module

This module contains several classes and methods that allow users to perform pairwise and multiple sequence alignment. Sequences can be aligned in both a single and multi-threaded fashion. BioJava implements the Needleman-Wunsch algorithm for optimal global alignments and the Smith and Waterman's algorithm for local alignments. The outputs of both local and global alignments are available in standard formats. In addition to these two algorithms, there is an implementation of Guan–Uberbacher algorithm which performs global sequence alignment very efficiently since it only uses linear memory.

For Multiple Sequence Alignment, any of the methods discussed above can be used to progressively perform a multiple sequence alignment.

ModFinder Module

Protein modifications are mapped onto the sequence and structure of ferredoxin I (PDB ID 1GAO). Two possible iron–sulfur clusters are shown on the protein sequence (3Fe–4S (F3S): Orange triangles/lines; 4Fe–4S (SF4): purple diamonds/lines). The 4Fe–4S cluster is displayed in the Jmol structure window above the sequence display.

The ModFinder module provides new methods to identify and classify protein modifications in protein 3D structures. Over 400 different types of protein modifications such as phosphorylation, glycosylation, disulfide bonds metal chelation etc. were collected and curated based on annotations in PSI-MOD, RESID and RCSB PDB. The module also provides an API for detecting pre-, co-, and post-translational protein modifications within protein structures. This module can also identify phosphorylation and print all pre-loaded modifications from a structure.

Amino Acid Properties Module

This module attempts to provide accurate physio-chemical properties of proteins. The properties that can calculated using this module are as follows:

- Molecular mass.
- Extinction coefficient.
- Instability index.
- Aliphatic index.
- Grand average of hydropathy.
- Isoelectric point.
- Amino acid composition.

The precise molecular weights for common isotopically labelled amino acids are included in this module. There also exists flexibility to define new amino acid molecules with their molecular weights using simple XML configuration files. This can be useful where the precise mass is of high importance such as mass spectrometry experiments.

Protein Disorder Module

The goal of this module is to provide users ways to find disorders in protein molecules. BioJava includes a Java implementation of the RONN predictor. The BioJava 3.0.5 makes use of Java's support for multithreading to improve performance by up to 3.2 times, on a modern quad-core machine, as compared to the legacy C implementation.

There are two ways to use this module:

- Using library function calls.
- Using command line.

Some features of this module include:

- Calculating the probability of disorder for every residue in a sequence.
- Calculating the probability of disorder for every residue in the sequence for all proteins from a FASTA input file.
- Get the disordered regions of the protein for a single protein sequence or for all the proteins from a FASTA input file.

Web Service Access Module

As per the current trends in bioinformatics, web based tools are gaining popularity. The web service module allows bioinformatics services to be accessed using REST protocols. Currently, two services are implemented: NCBI Blast through the Blast URLAPI (previously known as QBlast) and the HMMER web service.

Comparisons with other Alternatives

The need for customized software in the field of bioinformatics has been addressed by several groups and individuals. Similar to BioJava, open-source software projects such as BioPerl, BioPython, and BioRuby all provide tool-kits with multiple functionality that make it easier to create customized pipelines or analysis.

All of these APIs offer similar tools so on what criteria should one base their choice? For programmers who are experienced in only one of these languages, the choice is straightforward. However, for a well-rounded bioinformaticist who knows all of these languages and wants to choose the best language for a job, the choice can be made based on the following guidelines given by a software review done on the Bio* tool-kits.

In general, for small programs (<500 lines) that will be used by only an individual or small group, it is hard to beat Perl and BioPerl. These constraints probably cover the needs of 90 per cent of personal bioinformatics programming.

For beginners, and for writing larger programs in the Bio domain, especially those to be shared and supported by others, Python's clarity and brevity make it very attractive.

For those who might be leaning towards a career in bioinformatics and who want to learn only one language, Java has the widest general programming support, very good support in the Bio domain with BioJava, and is now the de facto language of business (the new COBOL, for better or worse).

Apart from these Bio* projects there is another project called STRAP which uses Java and aims for similar goals. The STRAP-toolbox, similar to BioJava is also a Java-toolkit for the design of Bioinformatics programs and scripts. The similarities and differences between BioJava and STRAP are as follows:

Similarities

- Both provide comprehensive collections of methods for protein sequences.
- Both are used by Java programmers to code bioinformatics algorithms.
- Both separate implementations and definitions by using java interfaces.
- Both are open source projects.
- Both can read and write many sequence file formats.

Differences

- BioJava is applicable to nucleotide and peptide sequences and can be applied for entire genomes. STRAP cannot cope with single sequences as long as an entire chromosome. Instead STRAP manipulates peptide sequences and 3D-structures of the size of single proteins. Nevertheless, it can hold a high number of sequences and structures in memory. STRAP is designed for protein sequences but can read coding nucleotide files, which are then translated to peptide sequences.
- STRAP is very fast since the graphical user interface must be highly responsive. BioJava is used where speed is less critical.
- BioJava is well designed in terms of type safety, ontology and object design. BioJava uses objects for sequences, annotations and sequence positions. Even single amino acids or nucleotides are object references. To enhance speed, STRAP avoids frequent object instantiations and invocation of non-final object-methods.

  - In BioJava peptide sequences and nucleotide sequences are lists of symbols. The symbols can be retrieved one after the other with an iterator or sub-sequences can be obtained. The advantages are that the entire sequence does not necessarily reside in memory and that programs are less susceptible to programming errors. Symbol objects are immutable elements of an alphabet. In STRAP however simple byte arrays are used for sequences and float arrays for coordinates. Besides speed the low memory consumption is an important advantage of basic data types. Classes in Strap expose internal data. Therefore, programmers might commit programming errors like manipulating byte arrays directly instead of using the setter methods. Another disadvantage is that no checks are performed in STRAP whether the characters in sequences are valid with respect to an underlying alphabet.
  - In BioJava sequence positions are realized by the class Location. Discontiguous Location objects are composed of several contiguous RangeLocation objects or PointLocation objects. For the class StrapProtein however, single residue positions are indicated by integer numbers between 0 and countResidues()-1. Multiple positions are given by boolean arrays. True at a given index means selected whereas false means not selected.
- BioJava throws exceptions when methods are invoked with invalid parameters. STRAP avoids the time-consuming creation of Throwable objects. Instead, errors in methods are indicated by the return values NaN, -1 or null. From the point of program design however Throwable objects are nicer.
- In BioJava a Sequence object is either a peptide sequence or a nucleotide sequence. A StrapProtein can hold both at the same time if a coding nucleotide sequence was read and translated into protein. Both, the nucleotide sequence and the peptide sequence are contained in the same StrapProtein object. The coding or non-coding regions can be changed and the peptide sequence alters accordingly.

Projects using BioJava

The following projects make use of BioJava:

- Metabolic Pathway Builder: Software suite dedicated to the exploration of connections among genes, proteins, reactions and metabolic pathways.
- DengueInfo: A Dengue genome information portal that uses BioJava in the middleware and talks to a biosql database.
- Dazzle: A BioJava based DAS server.
- BioSense: A plug-in for the InforSense Suite, an analytics software platform by IDBS that unitizes BioJava.

- Bioclipse: A free, open source, workbench for chemo- and bioinformatics with powerful editing and visualizing abilities for molecules, sequences, proteins, spectra, etc.
- PROMPT: A free, open source framework and application for the comparison and mapping of protein sets. Uses BioJava for handling most input data formats.
- Cytoscape: An open source bioinformatics software platform to visualize molecular interaction networks.
- BioWeka: An open source biological data mining application.
- Geneious: A molecular biology toolkit.
- MassSieve: An open source application to analyze mass spec proteomics data.
- Strap: A tool for multiple sequence alignment and sequence-based structure alignment.
- Jstacs: A Java framework for statistical analysis and classification of biological sequences.
- jLSTM "Long Short-Term Memory" for protein classification.
- LaJolla Structural alignment of RNA and proteins using an index structure for fast alignment of thousands of structures. Including an easy to use command line interface. Open source at Sourceforge.
- GenBeans: A rich client platform for bioinformatics primarily focused on molecular biology and sequence analysis.
- JEnsembl: A version-aware Java API to Ensembl data systems.
- MUSI: An integrated system to identify multiple specificity from very large peptide or nucleic acid data sets.
- Bioshell: A utility library for structural bioinformatics.

MOLECULAR EVOLUTIONARY GENETICS ANALYSIS

Molecular evolutionary genetics analysis (MEGA) is a bioinformatics tool used for genome analysis of molecular sequences to measure evolutionary distance for the construction of phylogenies. MEGA soiZDre package was designed at the Pennsylvania

State University lab under the supervision of Masatoshi Nei along with his postdoctoral fellow Koichiro Tamura and graduate student Sudhir Kumar who were responsible for writing the programming codes for MEGA. HLs tool was designed to extract valuable information from nucleotide or protein sequence for statistical assessment and biological data mining. Different versions of MEGA were released from time to time with better graphical user interface and enhanced features. For the purpose of comparative genome analysis MEGA performs the following:

- Sequence alignments.
- Evolutionary distance measurements.
- Phylogenetic tree building methods.
- Phylogenetic tree evaluation.
- Genes/Domains identification.
- Selection confirmation.
- Implementing sequence statistics.

MEGA Layout

MEGA extract valuable information from pairwise or multiple alignments of protein or DNA sequences by employing series of statistical techniques to determine specific physiognomies of nucleotide or proteins for the prediction of evolutionary relationships.

Centric Design of MEGA

User friendly and context dependent interface: MEGA displays a functional icon that gives users multiple options to choose from, relating to the current input data set that is being analyzed. It enables users to select specific features for computation depending on their designed analysis conditions and purpose of study. Thus making MEGA simple and user friendly.

Table: Major versions of MEGA.

MEGA Version	Release Date
1	1993
1.1	1994
2	2000
2.1	2001
3	2004
4	2006

4.1	2008
5	2011
5.1	2012
5.2	2013
6	2013
7	2016

Input Data Exploration

MEGA comprises of graphic modules for providing facilities like searching, editing, implementing and computing for the query data. It supports different formats (Microsoft Excel, CSV etc.) for exporting statistical results for graphical representations and further assessments. In addition functions to select or remove specific genes, domains, and species from primary data set have been possible by means of input data explorers. Besides features like selection of codon positions, codon translations, gaps deletions in alignments etc. have been incorporated in MEGA to for better analysis.

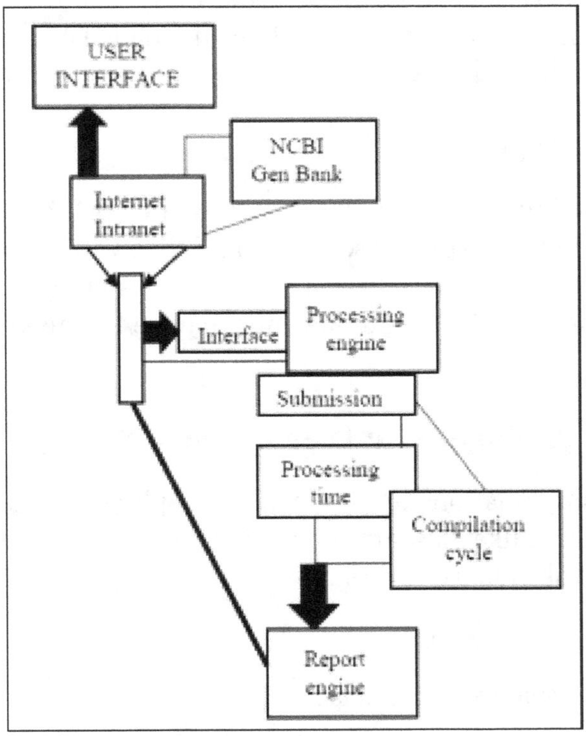

Mega working layout.

Output Explorers

MEGA contains many output explorers for graphical representation of the produced results. Such as tree explorer for displaying multiple tree representations, construction of consensus trees, sub-trees compression to identity superior relationships among sequences, printing trees as windows metafiles and TIFF files transforming trees into newick format, implementation of molecular clock to estimate evolutionary divergence for nodes in a tree etc. Exhibitions of pairwise distances with estimated standard errors is achieved by another output explorer called Distance matrix explorer that also possess sequence or taxa rearrangement facility by drag and drop.

Sequence Alignments and Data Assembly

Sequence data is directly imported into MEGA by alignment explorer that retrieve sequences from databases by means of in-built web browser that directly downloads the selected sequence files in FASTA or other formats. Besides alignment of any user selected sequence region could be edited either inserted or removed from the whole alignment set by means of alignment editor.

Elucidation and Transparency Results

Diverse computational and statistical analyses could be easily achieved in MEGA through intuitive GUI (Graphical user interface) which enables users to specify their requirements by choosing from multiple options manually i.e. user selection based decision to generate the desired output.

User Sessions Storage

Attribute of user sessions storage is incorporated in MEGA which allows users to save the current alignment explorer session to a file to which he or she could return later. HLs saving does not result in loss of alignment visual information and the user can resume the session any time and does not require the need to store data in text rich files.

Using MEGA for Phylogenetic Tree Construction

Computing evolutionary relationship by means of graphical representation of phylogenies using branching tree like diagram is one of the core applications of MEGA. Phylogenetic tree reveals evolutionary part of taxa.

Steps to create phylogenetic trees:

- Sequences retrieval from databases.
- MSA (multiple sequence alignments) using TCoffee ClustalW, MUSCLE etc.
- Tree estimation.
- Tree evaluation.
- Tree representation.

Step 1: Sequence Retrieval.

Building sequence alignments involves many steps including acquiring sequences from databanks using browsing facility. Homologous sequences are BLASTed by using either gene name or accession number or a query sequence etc. output containing similar sequence set is generated and displayed on the screen. From this set one can select either all or some sequences i.e. cutting and pasting the sequences from web browser or saving them into files for processing them for sequence alignment. To stream line this process MEGA now includes an integrated web browsing facility to facilitate sequence retrieval from data banks & BLAST search.

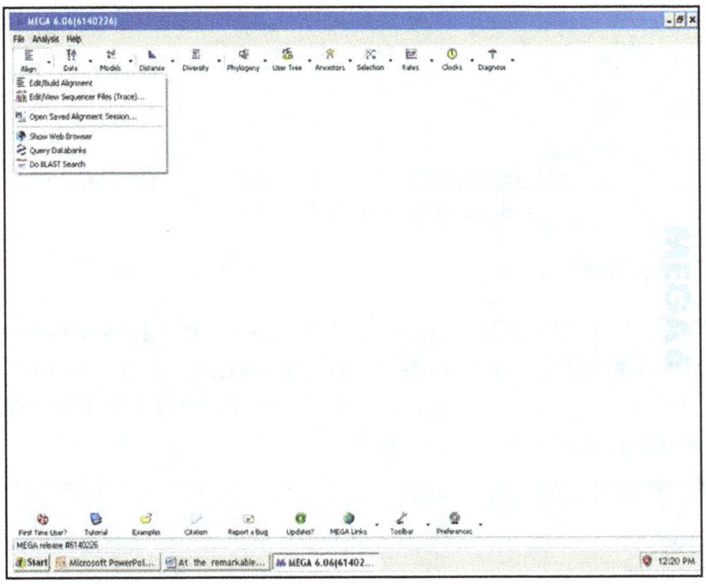

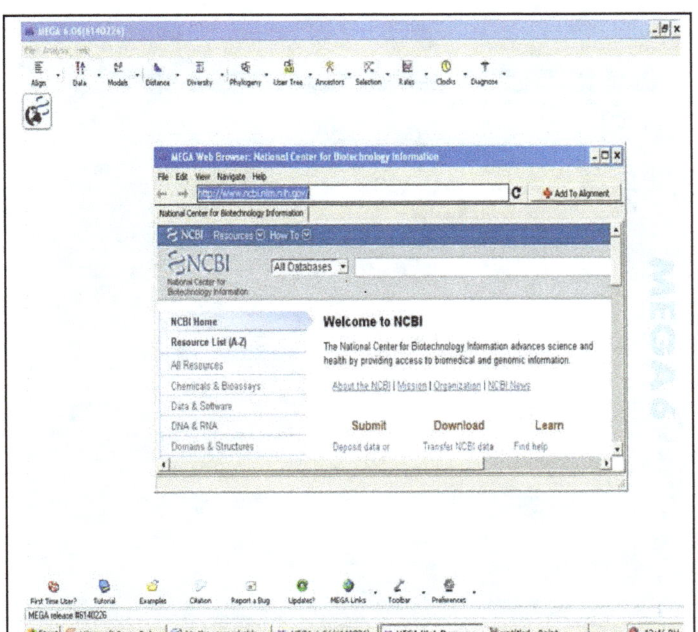

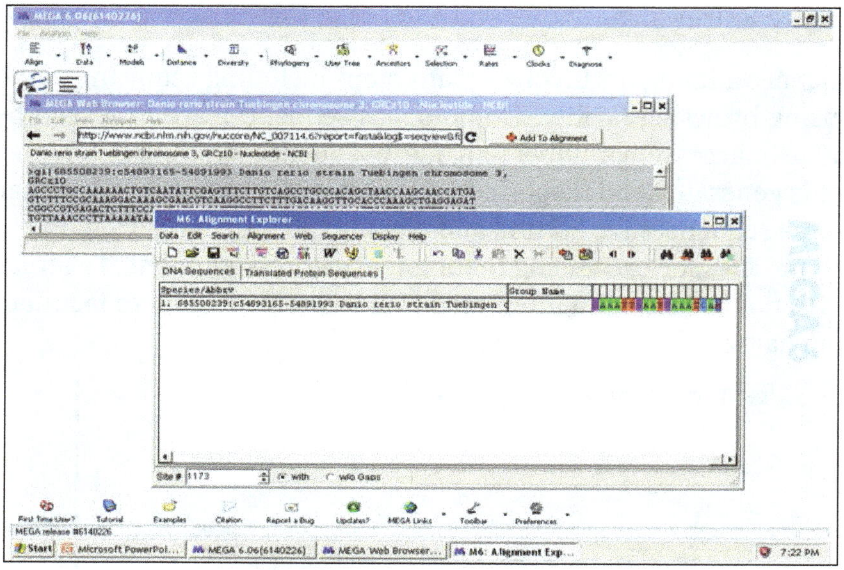

Sequences retrieval in MEGA.

Step 2: Sequence Alignment.

By selecting ADD TO ALIGNMENT option, MEGA exports the sequences automatically to the alignment explorer for producing multiple sequence alignments. Alignment explorer provides two approaches using ClustalW through which data could be displayed as DNA sequence and translated amino acid.

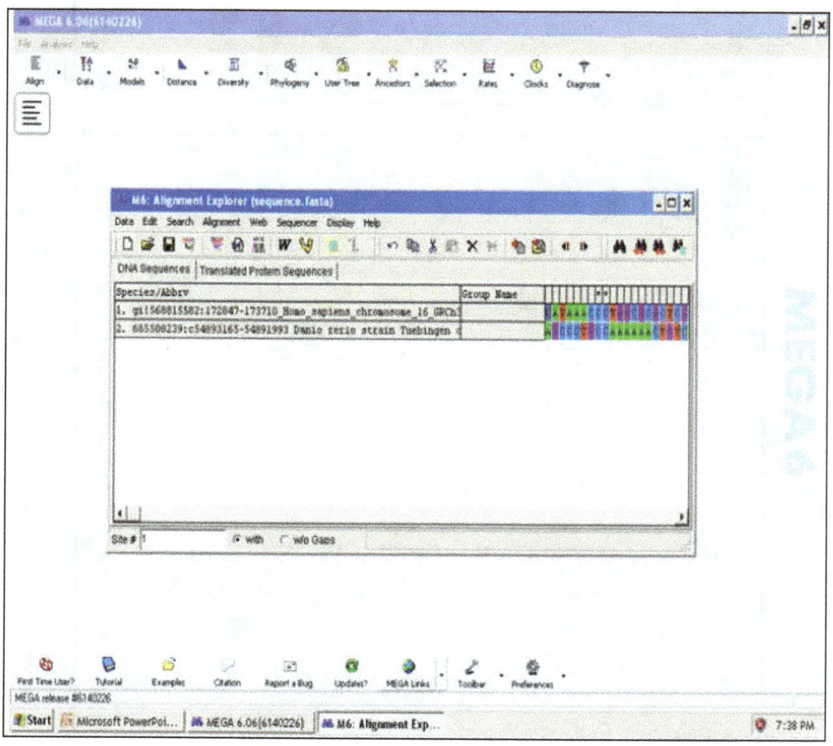

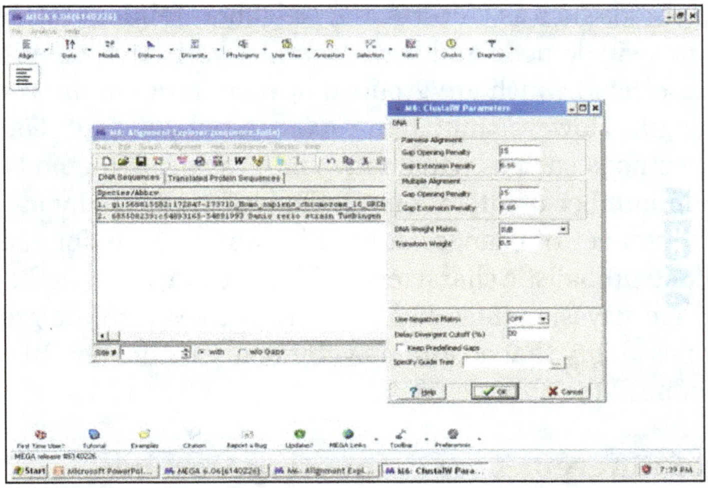

Showing the levels construction of data subset to remove error likely to occur during manual data editing.

Level 1: construction of data subset containing nearly any combination of sequences including groups, domains, and genes.

Level 2: Inclusion or exclusion of data with missing information or alignment gaps prior to analysis called as complete deletion or pair wise deletion.

Level 3: Data sub setting and transformation in MEGA is accomplished automatically by codon extraction from the selected data subsets and its translation if needed.

Step 3: Tree Estimation.

Here are several tree estimation methods given below.

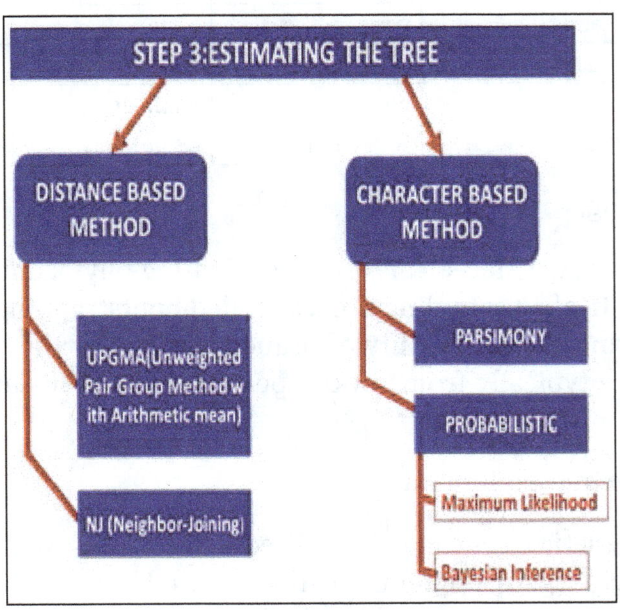

Tree estimation methods in MEGA.

Distance based methods such as UPGMA and Neighbor-Joining (NJ) begins with placing all the taxas in a single node and then separates with each repetition. In this way pair of nodes are selected which are grouped at each iteration in order to reduce the overall branch length. However mutation rates are not constant. On the other hand character based methods such as parsimony and probalistic methods selects the tree with the minimum number of alterations as the preferred tree by identifying and estimating the total number of changes at each informative site for each possible tree. Specifically speaking probalistic character based tree estimation methods such as maximum likelihood and bayesian inference finds set of trees with the greatest likelihood by assessing the possibility that a certain evolutionary model (eg. BLOSSUM or PAM matrices) has produced the observed data.

Comparison of different Tree Estimating Methods

Comparison of different tree estimating methods is mentioned above explaining which method is the most appropriate in terms of accuracy, time and convenience.

Table: Efficiency of major tree estimating methods.

Accuracy				
Bayesian inference				
Maximum likelihood (Decreasing accuracy)				
Maximum parsimony				
Neighbor joining				
Time and Convenience (In terms of time and convenience NJ is found to be the fastest)				
Data set	Neighbor Joining	Maximum parsimony	Maximum likelihood	Bayesian inference
Small	1 sec	3 sec	6 sec	-
Small	9 sec	10 min	1 h 34 min	29 min 40 sec
Small	1 sec	22 sec	3 min 29 sec	-
Large	86 sec	10 h 2 min	58 h	6 h 33 min

Step 4: Bootstrapping.

It is important to conduct statistical test for evaluating phylogenetic tree. Therefore MEGA runs a statistical re-sampling process called bootstrapping to check trees reliability by measuring the probability of branch recovery if the taxa were sampled again. Its values are typically from 1000 repeated calculations and values >70% is acceptable.

Step 5: Phylogenetic tree exploration.

Constructed phylogenetic trees can be visualized in numerous ways such as topologies without branch length e.g. Cladogram or with estimated branch length e.g. Phylogram or in linearized fashioned by means of tree explorer.

BIOINFORMATICS WORKFLOW MANAGEMENT SYSTEM

A bioinformatics workflow management system is a specialized form of workflow management system designed specifically to compose and execute a series of computational or data manipulation steps, or a workflow, that relate to bioinformatics.

There are currently many different workflow systems. Some have been developed more generally as scientific workflow systems for use by scientists from many different disciplines like astronomy and earth science. All such systems are based on an abstract representation of how a computation proceeds in the form of a directed graph, where each node represents a task to be executed and edges represent either data flow or execution dependencies between different tasks. Each system typically provides a visual front-end, allowing the user to build and modify complex applications with little or no programming expertise.

Example:

In alphabetical order, some examples of bioinformatics workflow management systems include:

- Anduril bioinformatics and image analysis.
- BioBIKE: A Web-based, programmable, integrated biological knowledge base.
- CSI NGS Portal: An online platform for automated NGS data analysis and sharing.
- Cuneiform: A functional workflow language for large-scale data analysis.
- Discovery Net: One of the earliest examples of a scientific workflow system, later commercialized as InforSense which was then acquired by IDBS.
- FireCloud: A free workflow, storage, and distributed compute SaaS workspace providing a repository of computational methods for running on the Cromwell engine. Built by Broad Institute on the Google Cloud Platform.
- Galaxy: Initially targeted at genomics.
- GenePattern: A powerful scientific workflow system that provides access to hundreds of genomic analysis tools.
- KNIME the Konstanz Information Miner.
- OnlineHPC Online workflow designer based on Taverna.
- Terra a workspace for bioinformatics, including a repository of public best practice methods, public data sets, and cloud and distributed computing facilities.

Terra builds on and integrates FireCloud. Created by the Broad Institute and Verily.

- UGENE provides a workflow management system that is installed on a local computer.
- VisTrails.

ARRAYTRACK

Microarray technology has been widely used in diverse fields, including pharmacology, toxicology and clinical medicine. An integrated bioinformatics infrastructure to facilitate data management and analysis is vital fully to realise the potential impacts of genomics on research and public health. The Food and Drug Administration (FDA) has developed the genomics tool, ArrayTrackTM, which has a rich set of functionalities to manage, analyse and interpret genomic data, with a focus on microarray data. Most functionality is applicable to biomarker development in biomedical research and personalised medicine using other molecular technologies.

ArrayTrack became the FDA's genomics tool to support the Voluntary Genomics Data Submission (VGDS) programme8 in early 2004. VGDS provides a novel mechanism outside normal regulatory interactions for sponsors and the FDA to develop expertise, tools and processes appropriate for regulatory interpretation of pharmacogenomics data. In addition to its broad use within the FDA in various research and regulation-related programmes, ArrayTrack is also freely available to the scientific community. Users can gain access to ArrayTrack either through the website or by requesting media for local installation. The ArrayTrack user base has grown steadily, and the tool has been adopted by several government agencies (e.g. the Environmental Protection Agency, Centers for Disease Control and Prevention and National Institutes of Health), academic institutions and private companies.

ArrayTrack was designed as a client-server system that provides: (i) A database server hosting microarray data, related study data and a collection of genomic functional information (eg gene annotations, pathways, gene ontology); (ii) A client application that integrates visualisation and analysis of microarray data with functional information; and (iii) A hierarchically modular structure that provides extensibility for other types of 'omics' data and data from other emerging molecular technologies (e.g. genome-wide association studies).

ArrayTrack comprises three major integrated components: (i) MicroarrayDB, a database that stores essential data associated with a microarray experiment, including raw gene expression data and information on samples, treatments and phenotypic observations; (ii) LIB, a series of libraries that contain functional information (eg gene annotations, protein function and pathways) from public sources; and (iii) TOOL, a set of algorithms that provide analysis capabilities for data visualisation, normalisation, significance analysis, clustering and classification. A user-friendly interface enables the selection of an analysis method from TOOL, applying the method to selected microarray data stored in MicroarrayDB and linking analysis results directly to associated functional annotations in LIB. Some key functionality and their applications are highlighted below and the full list of functions is available on the ArrayTrack website.

- Data import: SimpleTox, an internally developed Clinical Data Interchange Standard Consortium/Standard for Exchange of Non-clinical Data (CDISC/SEND)-compliant format, allows essential experiment summaries to be uploaded in a spreadsheet format. This consistent and easy-to-use format facilitates crossstudy analysis.
- Class comparison: ArrayTrack offers a rich set of methods for determining a list of differentially expressed genes (DEGs) by comparing two groups (e.g. treated versus control), such as the simple t-test, the Volcano plot and more advanced statistical approaches, such as false discovery rate (FDR) control and significance analysis of microarrays (SAM). In addition, the analysis of variance (ANOVA) is available for multi-group comparison.
- Class discovery: Principal component analysis (PCA), hierarchical cluster analysis (HCA) and K-means clustering are available for unsupervised class discovery and pattern identification. Such methods are powerful ways to investigate the grouping of samples in terms of their similarities in gene expression.
- Class prediction: For predictive model development, ArrayTrack offers supervised learning methods such as K-nearest neighbour (KNN), linear discriminant analysis (LDA) and support vector machines (SVMs). Model building is an important step towards applying microarray technology to diagnosis, prognosis and treatment outcome prediction.

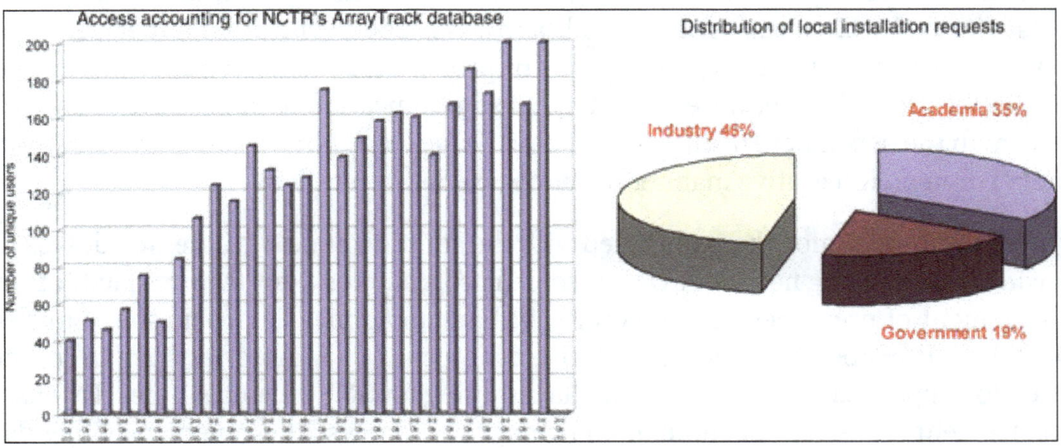

ArrayTrack user base statistics. The bar chart on the left represents the quarterly numbers of unique users who have accessed the ArrayTrack database hosted at the FDA's National Center for Toxicological Research (NCTR), excluding FDA users. The bar chart shows the steady growth of web users of ArrayTrack. The pie chart on the right displays the distribution among institutional type for 150 local ArrayTrack installations since 2004. Many industrial users come from pharmaceutical companies. The 'Academia' category includes public research-oriented organisations.

Importantly, most ArrayTrack functionalities are not limited to microarray data. For example, the information from ArrayTrack's extensive collection of libraries can be applied directly to data from nonmicroarray technologies, results collected from literature, or other sources. In addition, most statistical, classification and visualisation tools have been implemented as generic tools and can be readily used to analyse data where dependent and independent data are from other research fields. ArrayTrack provides a broad range of functionalities, allowing integration of functional information about genes, proteins, pathways and gene ontology with data from microarrays and other molecular technologies. Such integrative analysis capability furthers ArrayTrack's ability to support systems biology. ArrayTrack's diverse capabilities have been augmented recently, as described in the next section.

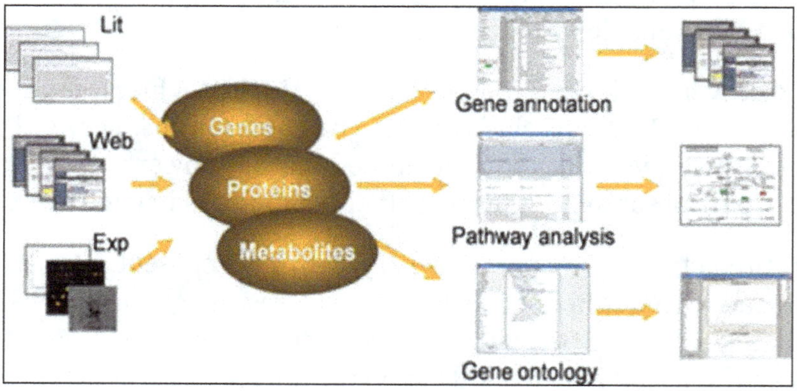

A workflow for data interpretation using the ArrayTrack library collection. The process starts with a list of genes, proteins and pathways. The lists can be either generated from in-house experiments (eg microarray or proteomics/metabolomics experiments) or collected from the literature, internet, or communication with peers. The user can easily map the lists independently to different ArrayTrack libraries (e.g. gene, pathway library) or in an integrative manner for biological interpretation.

ArrayTrack development has continued with the addition of new utilities to address the growing and changing needs of FDA's research programmes. New features facilitate the management of preprocessed proteomics and metabolomics data. The single nucleotide polymorphism (SNP) and quantitative trait locus (QTL) libraries have been integrated to support pathway analysis and data mining for SNP-related studies. Extensive enhancements have also been made to manage and analyse the genetic profiling data related to bacterial food-borne pathogens. These enhancements are depicted in figure in relation to ArrayTrack's core functionality.

Support for Proteomics and Metabolomics Data

Proteomics and metabolomics have grown steadily in importance in biomedical research, in parallel with microarrays. The integration of tri-omics data (i.e. genomics, proteomics and metabolomics data) has been a primary goal in systems biology for drug development and safety evaluation. To support this line of research and to review this type of data submitted by sponsors to the FDA through the VGDS – or, as it has been renamed, the Voluntary Exploratory Data Submissions (VXDS) program – ArrayTrack was previously modified to accommodate lists of proteins and metabolites. Additionally, a new systems biology function, called CommonPathway, was added that enables the examination of common pathways and functional categories (e.g. gene ontology terms) shared by different data types.

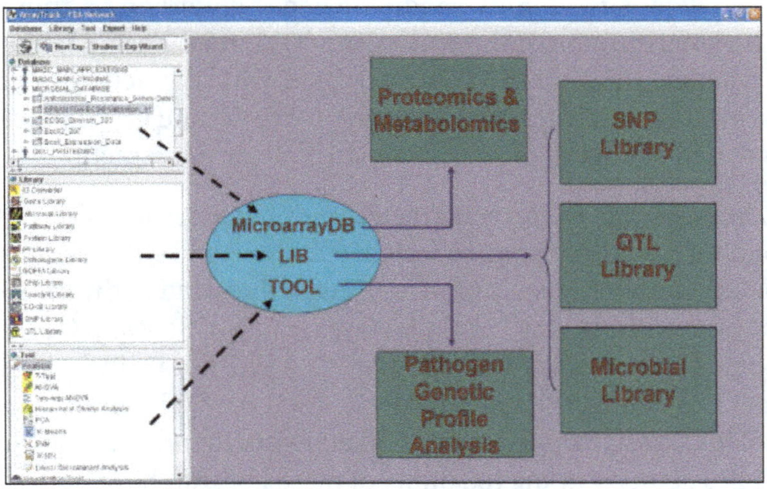

Recent enhancements to ArrayTrack.

ArrayTrack is now capable of analysing data from any mass spectrometry platform once the raw data have been processed for detection and quantification of peptides or metabolites – an important step in the data analysis workflow.15,16 New tools have been created to simplify the handling of proteomics data from the two most popular database search programs, Mascot and Sequest, for detection of peptides. The tools convert output files from these two programs into ArrayTrack readable files.

The same interpretation tools used for microarray data in ArrayTrack are extensible for proteomics and metabolomics data. Thus, by linking the results to gene, protein and pathway databases, researchers will be able to contextualise these results in the same way as gene expression experiments. Additionally, a unified interface helps to reduce the 'learning curve' associated with analysing new data types, giving researchers currently working on microarrays an incentive to move towards more integrated approaches that also encompass proteins and metabolites.

SNP and QTL Libraries

Recent advances in microarray-based genotyping techniques have enabled researchers rapidly to scan for known SNPs across complete genomes. An efficient data-mining strategy and a set of sophisticated tools are necessary better to understand and utilise the findings from genetic association studies.

One of the focuses in genetic association studies is to relate SNPs to genes and pathways in order to understand the underlying disease mechanisms.

ArrayTrack has already provided a gene –pathway discovery platform. By integrating the SNP library, which contains annotation summary information for SNPs and their mapped relationship to genes, ArrayTrack now provides an integrated SNP– gene–pathway analysis platform for SNP studies.

A QTL is a region of DNA that is associated with a particular phenotypic trait. A common use for QTL data is to identify candidate genes underlying a trait within one or more QTLs. The identification of the SNP–gene–QTL relationship is the basis of tests to determine whether the gene/SNP is associated with the aetiology of a disease in animal models or human studies. The integration of SNP and QTL libraries into ArrayTrack enables dynamic mining of such complex biological interactions.

SNP and QTL libraries have been constructed and incorporated into ArrayTrack. Data from several public repositories were collected in the SNP and QTL libraries and connected to other domain libraries (genes, proteins, metabolites and pathways) in ArrayTrack. Linking the data sets within ArrayTrack allows searching of SNP and QTL data, as well as their relationships to other biological molecules. The SNP library includes approximately 15 million human SNPs and their annotations, while the QTL library contains publicly available QTL data associated with specific phenotypes identified in mice, rats and humans. Case studies demonstrating the utility of these libraries have been reported.

Support for Microbial Pathogen Microarray Data

Food-borne pathogens are a leading cause of illness in the USA. High-throughput microarray technology provides an effective way to identify, characterise and obtain a nearly complete snapshot of the genetic traits of bacterial strains, such as their pathogenicity, virulence or antimicrobial resistance. Such genome-wide insight is necessary for accurate identification and discrimination of pathogens that may contaminate the food supply.

ArrayTrack has been extended to support microbial genomics research using microarrays ArrayTrack's libraries have been populated with bioinformatics data relating to bacterial pathogen species from the public domain. Data processing and visualisation tools have been enhanced with customised options to facilitate analysis of genetic profiling microarray data. Specifically, three new functions have been developed and are particularly effective for analysis of these microarray data: flagbased hierarchical clustering analysis (HCA), a flag concordance (FC) heat map and flag indicators in the mixed scatter plot (where 'flag' refers to a gene's presence or absence call). These functions are particularly relevant and effective for the identification and characterisation of bacterial pathogens using microarray genetic profiling data. The enhancements are displayed in figure. For example, the Microbial Library is the newest addition to ArrayTrack's collection of libraries. Currently, it holds 270,000 gene records from a total of 84 bacterial strains: 30 Escherichia coli, 39 Salmonella enterica, ten Shigella spp. and five Vibrio spp. Thus, as a starting point, the Microbial Library is focused on these four bacterial genera that are common food-borne pathogens. ArrayTrack also holds microbial pathway information from the Kyoto Encyclopaedia of Genes and Genomes (KEGG) for over 50 of these strains20 and gene ontology information for the E. coli K12 substrain MG1655. The gene annotations and sequences for the Microbial Library were downloaded from the National Center for Biotechnology Information (NCBI) website.

References

- Llabrés, M.; Rocha, J.; Rosselló, F.; Valiente, G. (2006). "On the Ancestral Compatibility of Two Phylogenetic Trees with Nested Taxa". Journal of Mathematical Biology. 53 (3): 340–364. arXiv:cs/0505086. doi:10.1007/s00285-006-0011-4. PMID 16823581
- Tools: genscript.com, Retrieved 16 march, 2020
- Slonczewski, Joan; John Watkins Foster (2009). Microbiology: An Evolving Science. New York: W.W. Norton & Co. ISBN 978-0-393-97857-5. OCLC 185042615
- Mega-core-of-phylogenetic-analysis-in-molecular-evolutionary-genetics-2329-9002-1000183: hilarispublisher.com, Retrieved 11 April, 2020
- Daugelaite J, O' Driscoll A, Sleator RD (2013). "An Overview of Multiple Sequence Alignments and Cloud Computing in Bioinformatics". ISRN Biomathematics. 2013: 1–14. doi:10.1155/2013/615630. ISSN 2090-7702

PERMISSIONS

All chapters in this book are published with permission under the Creative Commons Attribution Share Alike License or equivalent. Every chapter published in this book has been scrutinized by our experts. Their significance has been extensively debated. The topics covered herein carry significant information for a comprehensive understanding. They may even be implemented as practical applications or may be referred to as a beginning point for further studies.

We would like to thank the editorial team for lending their expertise to make the book truly unique. They have played a crucial role in the development of this book. Without their invaluable contributions this book wouldn't have been possible. They have made vital efforts to compile up to date information on the varied aspects of this subject to make this book a valuable addition to the collection of many professionals and students.

This book was conceptualized with the vision of imparting up-to-date and integrated information in this field. To ensure the same, a matchless editorial board was set up. Every individual on the board went through rigorous rounds of assessment to prove their worth. After which they invested a large part of their time researching and compiling the most relevant data for our readers.

The editorial board has been involved in producing this book since its inception. They have spent rigorous hours researching and exploring the diverse topics which have resulted in the successful publishing of this book. They have passed on their knowledge of decades through this book. To expedite this challenging task, the publisher supported the team at every step. A small team of assistant editors was also appointed to further simplify the editing procedure and attain best results for the readers.

Apart from the editorial board, the designing team has also invested a significant amount of their time in understanding the subject and creating the most relevant covers. They scrutinized every image to scout for the most suitable representation of the subject and create an appropriate cover for the book.

The publishing team has been an ardent support to the editorial, designing and production team. Their endless efforts to recruit the best for this project, has resulted in the accomplishment of this book. They are a veteran in the field of academics and their pool of knowledge is as vast as their experience in printing. Their expertise and guidance has proved useful at every step. Their uncompromising quality standards have made this book an exceptional effort. Their encouragement from time to time has been an inspiration for everyone.

The publisher and the editorial board hope that this book will prove to be a valuable piece of knowledge for students, practitioners and scholars across the globe.

INDEX

www.ingramcontent.com/pod-product-compliance
Lightning Source LLC
LaVergne TN
LVHW080614210926
848485LV00003B/156
9781639895106